Daylength and the Flowering of Plants

L. T. Evans
CSIRO, Canberra, Australia

W. A. Benjamin, Inc.
Menlo Park, California · Reading, Massachusetts
London · Amsterdam · Don Mills, Ontario · Sydney

Cover: Flowering and vegetative
plants of Japanese Morning Glory,
Pharbitis nil. The plant with
flowers on the front cover was
exposed to one long dark night
soon after the seedling emerged,
whereas the vegetative plant on
the back cover was grown in long
days from germination.

This book is in the
Benjamin Modular Program in Biology

Copyright © 1975 by W. A. Benjamin, Inc.
Philippines copyright 1975.

Printed in the United States of America. Published simultaneously in Canada.

Library of Congress Catalog Card No. 75-5308

ISBN 0-8053-1465-2
ABCDEFGHIJKL–AL–798765

W. A. Benjamin, Inc.
2727 Sand Hill Road
Menlo Park, California 94025

Preface

Like almost any area of biology that one enters by experiment, the physiology of flowering in plants soon leads one to wrestle with major questions of concern to all biologists, such as how plants respond and adapt to the environment around them, how they measure time, how their various organs intercommunicate and are integrated, or how the form of these organs is specified and controlled. All these questions have, in fact, been illuminated by experiments on the mechanisms underlying the control of flowering.

On the other hand, these underlying mechanisms are expressed with unsurpassed diversity in the flowering behavior of plants. Such diversity is important in permitting plants to adapt their reproductive cycles closely to the great variety of niches and climates in which they grow. So great is the range and so subtle the variety of response among them that it is all too easy to lose sight of the underlying principles. On the other hand, concentration only on the search for broad principles can lead us to forget the adaptive value of this variety of response.

I have tried in this book to navigate between these twin hazards and to communicate some of the flavor and excitement of research on the control of flowering, in the hope that students, agronomists, horticulturists, gardeners, and others may be tempted to explore it for themselves.

My particular thanks are due to Jane Vickers for typing the manuscript so carefully and to my colleagues Les Ballard and Rod King for commenting on it so vigorously. The permission of authors and journals to republish their figures is also gratefully acknowledged.

L. T. Evans

About the Author

Born in New Zealand, Lloyd Evans received his first degrees there, followed by a D. Phil. from Oxford University in England. After two years at the California Institute of Technology, in 1956 he joined the CSIRO Division of Plant Industry in Canberra, Australia, of which he is currently Chief. During 1963/64 he worked with S. B. Hendricks and H. A. Borthwick at Beltsville on several aspects of flowering physiology. This field has been his main research interest, but he is also concerned with photosynthesis and translocation, the limits to crop yield, and the ways in which plants respond to climatic factors.

Contents

Introduction

Gardeners and naturalists have long been
familiar with the marked seasonal regularity of
flowering time in many wild and cultivated
plants. The adaptive value of such behavior in
many environments has also been recognized, but
only in the last sixty years has its physiological basis begun to be unraveled.

Research on the physiology of flowering im-
pinges on many of the major problems of biology.
How organisms respond and adapt to their environments, how their growth is integrated and their
development regulated by hormones, and how morphogenesis can be switched from one kind of
organ to another, are questions we consider in
the following pages. The physiology of flowering
is, therefore, a microcosm of much of biology
and a fascinating one to explore by experiment.

This module begins with the discovery,
sixty years ago, of the widespread and powerful
control of flowering and plant development by
daylength. How this control is modified by temperature to provide a range of sensitive adaptive mechanisms in both wild and domesticated
plants is then considered. After noting the
virtues of various experimental systems we turn

to the sequence of processes in the leaves which perceive the daylength. These, when they lead to flowering, are referred to as photoperiodic induction and include the transformations of the pigment phytochrome in light and darkness, and their relation to other light reactions and to endogenous rhythms. As a result of photoperiodic induction, the leaves may export one or more substances which, if they reach the shoot apex in sufficient amount, switch it from leaf formation to flower initiation. Although we still do not know the identity of these floral stimuli or hormones, we do know when and how fast they move and that they can be transmitted from one plant to another by grafting. The evidence, as to whether there is one or more of them and whether leaves in non-inductive conditions export inhibitors of flowering, is discussed before we review the clues we have to the identity of the floral stimulus.

Finally, we turn to the nature of events at the shoot apex following the arrival of the floral stimulus, the processes referred to as floral evocation. The floral stimulus switches the shoot apex to a new pattern of activity, with a geometry very different from that of its previously repetitive leaf formation. However, the effects of daylength do not end there, and in many plants they continue to control floral morphogenesis, affecting whether or not the flower has petals, is male or female, sexual or apomictic, fully differentiated or reverting to vegetative growth. Daylength also controls many other aspects of plant growth, such as germination and tuber formation and dormancy, but the mechanisms which link daylength and flowering are of sufficient variety to emphasize the subtle adaptations of plants to their environment.

One

The Discovery and Adaptive
Role of Photoperiodism

Discovery

Following the invention of a practical incandescent lamp by Edison in 1879, experiments in
"electrohorticulture" by Bailey and others
showed that the flowering of several plants
could be accelerated by extending the natural
daylength, i.e., the period between sunrise and
sunset, with incandescent light. At the time,
this faster flowering was ascribed to a general
acceleration of growth by the additional light.
Klebs in Germany, however, came tantalizingly
close to realizing the significance of daylength
when he made plants of Sempervivum (house leek)
flower by exposing them to several days of continuous light. He concluded that the additional
light was acting catalytically rather than
nutritionally, but he left the matter there.
Experiments with four plants whose response
to daylength was the opposite of that of Sempervivum led to the discovery of photoperiodism.
Julien Tournois of Paris was puzzled as to why
plants of hops (Humulus japonicus) and hemp

3

(Cannabis sativa) should flower so precociously
in greenhouses when sown in winter. Temperature,
humidity, and seed origin were eliminated as
causal factors in his early experiments. Then,
in 1912, he exposed groups of plants, from
sowing, to either natural spring daylengths, or
continuous light, or days shortened to six hours
by placing covers over them. The plants in
this last treatment grew most slowly but flowered most rapidly. Tournois initially concluded
that hops and hemp flowered faster when they
received a smaller quantity of light, but in
subsequent experiments he found that reduced
light intensity had little effect on their
flowering. Consequently in his last paper,
published in 1914, Tournois concluded that the
precocious flowering of hops was caused by
exposure to short periods of daily illumination
or perhaps by the longer nights that accompanied
them. His death in the war a few months later
prevented him from developing this concept
further, and it was left to two Americans,
W. W. Garner and H. A. Allard, to establish daylength as a major environmental factor in the
control of flowering.

   Despite considerable early interest in the
United States in soybeans as a potential oilseed
crop, agronomists experienced difficulty in
adapting them to higher latitudes where they
would flower too late in the autumn to develop
worthwhile yields. In their experiments at Arlington, Virginia, Garner and Allard were
puzzled to find that Biloxi soybeans sown in the
field at intervals throughout spring and summer
all tended to flower at the same time. A seasonal factor was clearly involved, but extensive
experiments on the effects of light intensity
and spectral composition excluded them as the
causal agents. At the same time, Garner and
Allard were also trying to find out why a particular mutant of tobacco did not flower but

continued to grow throughout summer in the
field, hence its name of Maryland Mammoth. Old
plants of this type when moved to a greenhouse
in winter, or seedlings sown there, soon flowered. At first this was thought to be due to
poor nutrition, low light intensity, or the
shock of transplanting, but further experiments
eliminated these factors as the causative
agents. Garner and Allard concluded that the
only remaining seasonal phenomenon that could be
a factor was change in the relative length of
day and night. They did not pin much hope on
this conclusion, and made their test in the
simplest possible way. Starting on July 10,
1918, Allard placed a box of Peking soybeans and
three persistently vegetative Mammoth tobacco
plants into a light-tight "doghouse" (Figure 1)
for seventeen hours each night; comparable

Figure 1. Allard's "doghouse," used to impose
short days on soybeans and tobacco in 1918.

plants remained in the full summer daylength.
Those plants which received full summer daylength remained vegetative, but all the soybean
and tobacco plants exposed to short days and
long nights promptly flowered.

Given the extreme reluctance of Tournois,
Garner and Allard, and their contemporaries to
consider that daylength could be the environmental component so decisively in control of flowering, their choice of four short day plants--
hops, hemp, Mammoth tobacco, and soybeans--was
indeed fortunate. Whereas the acceleration of
flowering in plants like Sempervivum under extended daylengths seemed to be due simply to the
additional light, so dramatic a response to
short days was less readily explained in this way,
especially when reduced light intensities did
not affect flowering. Tournois was forced to the
conclusion that daylength was involved after
doing his crucial experiment, whereas Garner and
Allard reached it beforehand, although with
little hope.

One of the consequences of their experiments was the realization that growth and reproductive development are to a large degree independent of one another in plants, certainly more
so than in animals. Under favorable daylengths,
plants may flower precociously before they have
grown to any extent, as in Tournois' hops. In
unfavorable daylengths, they may grow to enormous size but never flower, as in the case of
Garner and Allard's Mammoth tobacco.

The four plants which played so decisive a
role in the recognition of daylength as a major
controlling factor were all economic plants
being studied with the objective of better agri-
cultural adaptation. The discovery of photoperiodism has led to their better adaptation to
a wider range of conditions. Soybeans, for example, were subsequently selected for daylength

responses which allowed them to flower early
even in the latitudes of the most northern
states, thus paving the way for their development as one of the major crops of the United
States.

However, Allard was a keen naturalist with
an extraordinarily wide knowledge of the behavior of plants, insects, and birds, and he and
Garner included many wild plants in their later
experiments on the effects of daylength. For
these experiments (at Arlington), they built a
series of ventilated but light-proof pavilions
in which they could shorten or extend the natural daylength by various amounts. Many plants
which flowered naturally in the autumn were
found to require exposure to days shorter than
a certain critical length before they would
flower. These they called short day plants
(SDP). The critical daylength varied consider-
ably between species, from about sixteen hours
to less than twelve hours. Some plants had an
absolute requirement for short days, like Mammoth tobacco, but others merely flowered much
faster in short days and have come to be referred to as quantitative short day plants. For
both kinds, however, the shorter the daylength,
down to about eight hours, the faster the
flowering.

On the other hand, many spring and summer
flowering plants such as radish, lettuce, and
Hibiscus proved to be either absolute or quantitative long day plants (LDP), flowering faster
the longer the daylength up to sixteen hours or
so. For the absolute LDP there was a critical
daylength, ranging from eight to twelve hours,
below which they would not flower at all.

A few plants, both wild and cultivated,
were found to be indifferent to daylength, such
as Job's Tears grass, Coix (see Figure 2, next
page). Subsequent experiments by Allard and

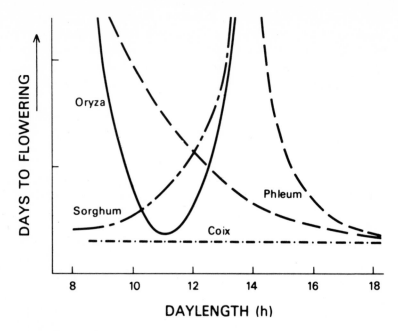

Figure 2. Response to daylength for flowering in several grasses and cereals. Coix (Job's Tears) was found by Garner and Allard to be indifferent to daylength. Phleum (Timothy) was shown by Allard and Evans to be a long day plant whose strains differed in their critical daylength. Sorghum roxburghii is a short day plant, while some strains of Oryza (rice) are intermediate day plants.

others revealed the existence of a number of plants, such as the wild climber Mikania scandens and Saccharum spontaneum, the wild sugar cane, which would flower in neither short nor long days but only at intermediate daylengths (e.g., the rice in Figure 2). These were called intermediate day plants. In their 1923 paper Garner and Allard also observed that not only time to flowering was responsive to daylength but also the kinds of flowers formed by violets, entry into dormancy by _Hibiscus_, the formation of tubers in potatoes, yams, and Jerusalem artichoke, pigmentation in _Poinsettia_, abscission and leaf fall in sumac and the tulip tree, and many other aspects of plant development. In their 1920 paper they speculated that comparable control of behavior by daylength, which they called photoperiodism, might apply to lower plants, insects, and birds. In fact, responses to daylength have been found in all but the lowest organisms, and man himself is not immune to their influence. Thus, these simple experiments of sixty years ago have had a profound impact on our understanding of environmental influences on the behavior, as well as the growth, of most organisms. This kinship of response, shared by man, adds to the fascination of experimental work on photoperiodism.

Just why it was so difficult for Tournois, Garner, and Allard to accept the controlling role of daylength is difficult to understand now that we are aware of its pervasive influence. As the most regular, indeed entirely predictable, seasonal factor, it seems a natural candidate as the environmental signal with least noise. Perhaps the real stumbling block was the implication that plants could in some way measure time accurately and by some mechanism that could compensate for variations in temperature.

Responses to Daylength and Temperature : Variations on a Theme

One of the problems posed by the use of daylength as an environmental signal is how to differentiate the equinoctial days of spring from those of autumn. One solution would be the

capacity to respond only to daylengths which are
progressively increasing or decreasing. In one
sense at least there are two groups of plants
which do this. Among autumn-flowering plants
there are some--such as several species of the
succulents Bryophyllum and Kalanchoë first
worked on by Dostal in Czechoslovakia and
Resende in Portugal, and the night-blooming jas-
mine Cestrum nocturnum investigated by Sachs in
California--that require exposure first to long
days and then to short days before they initiate
flowering (LSDP). Similarly, among spring-
flowering plants there are some--such as Scabi-
osa succisa used by Chouard in Paris, Campanula
medium by Wellensiek in Holland, and the strain
of white clover (Trifolium repens) investigated
by Thomas in New Zealand--that require the opposite sequence of daylengths, first short days
and then long ones (SLDP).

An alternative solution to the problem of
differentiating spring and autumn daylengths
would be to combine information on temperature
with that on daylength. Garner and Allard discussed the difference between spring and autumn
flowering quite fully in their 1923 paper and
ascribed it to the higher temperatures which
prevail at a given daylength in autumn, because
of the slow temperature rise in spring. In their
view most SDP do not flower in spring because,
by the time temperatures have risen enough for
growth to occur, the daylength has become too
long. Thus they found that autumn-flowering SDP,
like Peking soybeans, would flower in spring
when grown in greenhouses at temperatures above
those in the field.

However, temperature can also modify the
response to daylength. With warm nights, for
example, strawberries, oranges, and Japanese
morning glory (Pharbitis nil) are strict short
day plants, but at low temperatures they flower

in any daylength. Similarly, the long-short day
plants <u>Bryophyllum</u> <u>daigremontianum</u> and <u>Cestrum</u>
<u>nocturnum</u> will flower in long days if the nights
are cool, $12°C$ or less. By contrast, <u>Bouvardia</u>
<u>humboldtii</u> flowers in any daylength at high tem-
peratures, in long days only at moderate temperatures, and in no daylength at low temperatures.

Vernalization

In many long day plants flowering occurs only on
shoots previously exposed for a long period to
temperatures below $10°C$; this effectively con-
fines flowering to spring and early summer. This
subjection to prolonged low temperatures in the
seed or seedling stage or in overwintering ro-
settes and buds is referred to as vernalization
because it brings these plants to the condition
they reach at the onset of spring, namely responsiveness to long days. Some plants, however,
require only vernalization, i.e., prolonged exposure to cold, in order to initiate flowering,
being then indifferent to daylength. The perception of and response to low temperatures is
localized in the meristematic (i.e., dividing)
cells of embryos, growing points, and buds.
Thus, the branches of deciduous trees or under-
ground seeds and bulbs can sense the approaching
end of winter even in the absence of leaves. The
nature of the changes induced in meristematic
cells by prolonged low temperatures is still not
clear. Whatever they are, once vernalization is
complete, these changes can be conserved through
many generations of cell division at temperatures too high for vernalization to proceed.
This is indicated by experiments with varieties
of winter cereals and perennial grasses that require vernalization to be followed by long days
before they initiate flowering. Once vernalized,

these grasses and cereals can be grown in warm
short days for a year or more without showing
any progress towards flowering, but they do
flower as soon as they are placed in long days.
Their earlier vernalization is thus "remem-
bered," provided the exposure to cold has been
long enough for vernalization to reach
completion.

Vernalizable plants differ greatly in their
requirements for exposure to low temperatures,
most of which range between one and sixteen
weeks below 10°C. Chervil (Anthriscus cerefo-
lium) is known to respond to as little as one
day of vernalization. At the other extreme is
Geum urbanum whose axillary buds flower after
vernalization for two or three months, but whose
terminal buds form flowers only after vernalization for a full year. By this means the plant
can flower in spring each year on its axillaries
while maintaining perenniality in a terminal
vegetative shoot apex whose requirement for vernalization is more prolonged than it ever expe-
riences in the field.

Photoperiodism and Adaptation

The preceding discussion has touched on only a
small part of the enormous variety in the flowering response of plants to daylength, temperature, and their interactions. Chouard and his
colleagues at Paris have been particularly
active in exploring this variety, as may be
gathered from his 1960 review. Highly complex
schemes of behavioral classification have been
developed, and one of them is given in
Salisbury's (1963) book. This rich diversity of
response to daylength has often caused despair
in biologists in search of unexceptionable generalizations, but it is surely to be expected

when we are dealing with a powerful adaptive
mechanism under considerable evolutionary pres-
sure as climates and habitats change and plants
disperse. No comprehensive estimate has been
made of the proportion of wild plants which respond to daylength and vernalization, but it is
probably quite high in many environments.

Kangaroo grass, Themeda australis, is
almost the only species which occurs throughout
Australia, from Perth to Sydney and from New
Guinea to Tasmania. It is a perennial grass
which probably entered Australia from the
tropics. The northern, tropical populations of
this grass are strict short day plants, a response which results in flowering in autumn at
the end of the wet monsoon, a most favorable time.
Populations from middle-latitude coastal areas
are intermediate day plants, while those from
southern Australia are mostly long day plants
which flower towards the end of the winter-
spring period of rainfall. At the higher latitudes their requirement for long days is often
absolute, and the southernmost populations may
also have a strong requirement for vernalization, a rather remarkable adaptation for a plant
of tropical origin. Of all the populations examined the only ones indifferent to daylength
were those from the dry interior, where rain is
so sporadic that it is presumably advantageous
for flowering to occur at whatever season rain
falls.

The cocklebur, Xanthium strumarium, is a
wild plant mentioned frequently in the pages
that follow because of the requirement by one
strain for exposure to only one short day. All
strains are short day plants, but Ray and
Alexander found the critical dark period length
to vary with latitude from as little as seven
and one-half hours for populations near Montreal
to more than ten hours in Texas. In Mexico,

however, there is considerable variation in
critical night length even among populations at
one latitude, which McMillan associates with
differences in climate and topography. On the
basis of similarity of critical night length
McMillan (1973) suggests that the Xanthium populations of Tahiti, Oahu, and other Pacific
Islands are derived from Mexican coastal populations rather than from California. But as we
have seen with Themeda, daylength response
readily adapts to new environments, and similarity of daylength response is probably a poor
basis on which to draw conclusions about the
origin of plants.

Nor is it always easy to see in what way a
particular response to daylength is adaptive.
Daylengths in the tropics could satisfy the
needs of either short or long day plants, yet
hardly any of the latter are found there. A possible explanation is that the high night temperatures for about twelve hours of darkness are
favorable for many SDP but tend to be inhibitory
for LDP.

Domesticated Plants

With domesticated plants, daylength or vernalization requirements may restrict the range of
cultivation desired by man, as in the case of
early soybean growing in the United States.
Moreover, man assumes control of the life cycle
via time of planting and various cultural prac-
tices. As a result, the environmental requirements for flowering in cultivated plants have
tended to become muted by selection as the
plants have been domesticated and spread more
widely. Modern cereals are less demanding than
their wild progenitors in this respect, although
more demanding of high soil fertility. For

example, Katayama (1971) found that of nine species and 163 strains of rice, nearly all wild populations were sensitive to daylength but only one quarter of the cultivars. Many horticultural plants have also had their sensitivity to daylength reduced in the course of selection. Likewise, tuberization in wild Andean potatoes has a strict requirement for short days which has been largely eliminated in modern cultivars. One of the explicit objectives of the plant breeding programs at the International Rice Research Institute and for the Mexican wheats has been indifference to daylength, particularly to hasten the spread of new varieties around the world.

Nevertheless, daylength and vernalization requirements may still be of advantage to crops in some environments. Many Canadian spring wheats retain an absolute requirement for long days which may serve to delay inflorescence development until all risk of frost has passed. Likewise, the floating rices that grow in the monsoonally flooded rivers of Southeast Asia have a strict short day requirement which may prevent flowering and seed development until the floods subside. On the other hand, there are crops like sugar beet and sugar cane in which flowering depresses yield. Selection in these cases is for requirements which suppress flowering in the field but not, of course, under the conditions in which new cultivars are bred. Too much emphasis may have been given to selection for indifference to vernalization and daylength as an aid to wider adaptation. Close adaptation to particular environments can be an important component of high crop yields, particularly through mechanisms that delay the initiation of flowering until a deep root system and an effective photosynthetic canopy are established. For this purpose, a prolonged juvenile phase or pronounced vernalization and

photoperiodic requirements may be as valuable in
a crop as in a natural plant community.

Two

Experimental Plants and Systems

Decisive experiments on flowering are not easy to do on plants with a leaky (i.e., quantitative) response to daylength, and they are tiresome, to say the least, on plants that require an extended period in the appropriate daylength for photoperiodic induction. Plants with absolute requirements for long or short days give control groups which are entirely vegetative, and using them makes it easier to detect even the slightest response to an inductive daylength. It is in the nature of experiments that one is often exploring progressively more marginal responses as treatment differences are more closely analyzed, and an absolute daylength requirement for flowering is often advantageous in such experimental systems.

The progression towards experiments with a single inductive cycle is a case in point. To anticipate a little, if we want to measure the speed at which the floral stimulus moves out of an induced leaf or the timing of floral morphogenesis in relation to when that stimulus reaches the shoot apex, we can do so effectively only after "one-shot" photoperiodic induction.

There is another advantage to single-cycle
plants of which most people who have done complex flowering experiments are acutely aware.
Such experiments often require access to the
plants at all hours of the day and night. For
one inductive cycle, and for the days which pre-
cede and follow it, such a schedule is toler-
able, but over prolonged periods of induction
it is a flowering physiologist's nightmare, the
prospect of which may persuade one that the experiment is not worth doing.

The demand for responsiveness often goes
beyond single-cycle induction to the requirement
that this be achieved by exposure of only one
leaf or pair of leaves. This is essential where
complex light intensity and spectral quality
treatments are involved or where the export of
the floral stimulus from the leaf is to be
followed.

Not many plants match these stringent requirements, but there are enough to reveal a
considerable diversity of response.

(a) <u>Short day plants</u>. Among short day plants
the first, and for a long time the only plant to
match these requirements, was a Chicago strain
of the cocklebur <u>Xanthium strumarium</u>, often referred to as <u>X</u>. <u>pensylvanicum</u>. Hamner, in 1938,
noticed that cocklebur plants flowered after
accidental exposure to one long night and thereafter <u>Xanthium</u> became the <u>Drosophila</u> of photoperiodism. Indeed, Salisbury's book *The
Flowering Process* is written around it.

In 1955 Imamura and Takimoto introduced the
Violet strain of Japanese Morning Glory, <u>Phar-
bitis nil</u>, which has proved to be a superb experimental plant. One advantage it has over
<u>Xanthium</u> is that abundant flowering can be in-
duced by a single short day given only a few
days after the seeds have been sown, which

greatly increases the rate of experimentation.

Chenopodium rubrum, from latitude 62° 46' north, introduced by Cumming in 1959, provides an even simpler experimental system in that up to 150 seedlings can be grown on filter paper in a Petri dish, induced to flower by exposure to a single short day as soon as the cotyledons have emerged two to three days after sowing, and the flowering response assessed before the seedlings need to be transplanted from the dish. It is therefore extremely convenient for experiments on endogenous rhythms and the spectral composition of light but not so suitable for experiments on, for example, the movement of the floral stimulus. Both Pharbitis and Chenopodium require close scheduling and control of temperature and light before and after the inductive day for reproduceable results to be obtained.

Three duckweeds, strains of Lemna perpusilla, L. paucicostata, and Wolffia microscopica, are also responsive to one short day. Grown in flasks on various solutions, they are particularly suitable for exploring the effects of various nutrients and inhibitors of flower induction, as in many elegant experiments by Hillman.

(b) Long day plants. For many years our understanding of the photoperiodic reactions of LDP lagged behind that of SDP for lack of an experimental plant with an absolute requirement for only one long day. I therefore thought it worth investing a great deal of time and effort to find such a plant; in 1956, after an extensive search, I found one, darnel grass or Lolium temulentum. Only a small part of one leaf needs to be exposed to the one long day, and the species has proved to be very suitable for a wide range of work. Its chief disadvantage is

that it needs to be grown for five weeks in
short days before it responds to a single long
day. In this respect the pimpernel, Anagallis
arvensis, is better in that it can be used in
the seedling stage or as rooted cuttings, although it has other disadvantages. A strain of
spinach and of the duckweed Lemna gibba are
partially inducible by a long day but have not
proved to be very suitable experimental plants.

Four other plants can be induced to flower
by exposure to a single long day, but all of
them eventually flower in short days, especially
at high light intensities. Thus, their flowering
appears to be controlled by assimilatory as well
as by strictly photoperiodic reactions. Dill
(Anethum graveolens) was the first of these to
be used, the others being cultivars of mustard
(Sinapis alba), rape (Brassica campestris), and
barley (Hordeum vulgare). Of these the most extensively used has been the white mustard with
which Bernier and his colleagues at Liège in
Belgium have explored the reactions occurring
at the shoot apex following the long day
exposure.

(c) Single cycle induction - is it atypical?
Much of the current research on photoperiodism
is carried out on these "single-cycle" plants,
and the question naturally arises as to how
representative they are of the great majority
of less sensitive plants. To answer this we
need to know the basis of their requirement for
only a single inductive cycle. Hillman suggested
that they might be less subject to inhibition of
flowering by non-inductive daylengths, but this
is unlikely for at least several of them. Jacobs
(1972) associates their greater responsiveness
with a shorter interval between the formation of
successive leaf primordia. This interval, called
the plastochron, may be less than two days in

Xanthium which requires only one SD, whereas it is four to eight days in Perilla which requires seven to nine SD for induction. Jacobs cites several other examples to support his view, but there are also quite striking exceptions. Anagallis, for example, with a plastochron of almost four days, requires only one LD for induction. And in Lolium the requisite number of LD falls from more than four in young plants to one in older plants without any change in the length of the plastochron. Nevertheless, there may be a particular stage in the interval between the initiation of one leaf primordium and that of the next which is particularly sensitive to induction, which could account for Jacobs' observation that highly responsive plants tend to have short plastochrons.

Many single-cycle plants also have in common an origin from high latitudes. The Chenopodium rubrum strain derives from latitude 62° 46' in Canada, the darnel also from Canada, the cocklebur from Chicago, and Zuiho rice from Japan. In Polygonum thunbergii, Sawamura found that the more northern the origin of the strain, the fewer were the short days required for flower induction.

Nevertheless, these single-cycle plants encompass a substantial range of response, which they share with less sensitive experimental plants. Plants requiring more than one inductive day may sometimes have advantages as experimental material. Only with them is it possible to ask the questions "Where are the effects of single inductive days added up, in the leaves or at the shoot apex?" and "Are their effects quickly dissipated or in some way conserved?" Moreover, apart from Xanthium, the single-cycle plants have not proved to be as suitable for experiments on graft transmission of the floral stimulus as plants like tobacco and Perilla,

and such grafting experiments continue to play
a crucial role in shaping our interpretations
of the induction of flowering.

Details of how to handle most of these
plants experimentally are given by those most
familiar with them in a book entitled *The Induction of Flowering* (Evans, 1969).

The Measurement of Flowering Response

In many early experiments investigators were
concerned simply with whether or not plants
flowered or with the percentage of plants that
eventually flowered in a treatment group. When
differential photoperiodic treatments were applied from sowing, their relative effectiveness
could be gauged by the time that elapsed before
flowering occurred. However, effects of temperature and daylength on growth as well as those
on flower induction often confounded such measurements. To a degree this can be avoided by
using the formation of nodes or leaves on the
main stem as the time scale, instead of days.
Groups of plants can then be compared in terms
of the lowest nodes at which they form flowers.
There is always the possibility, however, that
a treatment might affect growth but not the rate
of flowering, thereby also changing the node of
first flowering.
      The simple measure of the number of flowers
formed must also be interpreted with care. In
many plants more effective daylength treatments
increase the number of flowers formed, but the
relation between stimulus and response depends
more on inflorescence structure than on the
kinetics of induction. In the succulent Kalanchoë blossfeldiana, for example, additional
short days increase the order of branching in

the inflorescence, so that flower number increases logarithmically with increase in the number of SD (Figure 3b). With many plants that bear their flowers in the axils of leaves and in which the axillary buds are only briefly susceptible to conversion to flowers, such as the SDP soybean and the LDP Anagallis, additional inductive days simply result in a proportional increase in the number of nodes bearing flowers (Figure 3a). In plants like Pharbitis, however, the effectiveness of short day induction may increase to the point where the terminal growing point of the shoot is also converted to a flower. No additional flowers can then be formed, and flower number approaches an asymptote. Indeed, in temperate cereals like wheat, the more effective the long day induction, the sooner the

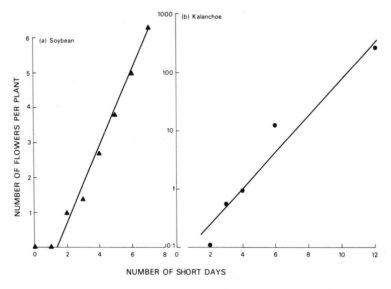

Figure 3. The effect of number of short days on the number of flowers per plant in (a) soybean (data of Hamner, 1940), (b) Kalanchoë (data of Schwabe, 1956).

terminal spikelet is formed, and the smaller is
the total number of spikelets, and the yield
potential.

The rate of development of the inflorescence after exposure to inductive daylength
treatments is a very sensitive measure of the
effectiveness of induction in some plants such
as Xanthium and Lolium. After exposure to a
short or long day respectively, the plants are
returned to non-inductive daylengths at a stan-
dard temperature, and the stage of development
of the induced inflorescences is determined by
dissection at intervals thereafter. For Xanthium, Salisbury has scored a series of floral
stages (see Figure 4, next page) that give a
linear plot with time. A comparable system of
morphological stages can be used in Lolium experiments, but a more objective measure is given
by the length of the shoot apex which increases
exponentially with time at a rate dependent on
the effectiveness of floral induction (see
Figure 5, page 26).

The most appropriate quantitative measures
of the flowering response thus depend on the
plants used: percent of plants flowering for
Chenopodium and Lemna, number of nodes with
flowers for soybean, number of flowers or percent terminal flowering for Pharbitis, and rate
of inflorescence development in Xanthium and
Lolium. Whatever the measure, photoperiodic induction is clearly a quantitative process, not
merely a "flower on/flower off" switch.

What Organ Perceives the Daylength?

The shoot apex is where the ultimate response
to vernalization and daylength occurs because
the floral primordia are formed there, and it is
also the site of the primary vernalization

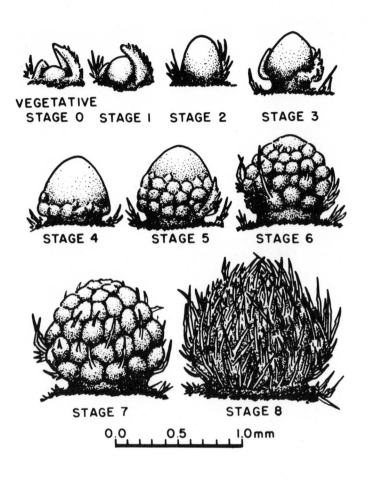

VEGETATIVE
STAGE 0   STAGE 1   STAGE 2   STAGE 3

STAGE 4   STAGE 5   STAGE 6

STAGE 7   STAGE 8

0.0   0.5   1.0mm

Figure 4. Stages in the development of the terminal male inflorescence of <u>Xanthium strumarium</u> (Salisbury, 1963).

reaction. The photoperiodic processes, however, occur in the leaves, as was first proved by Knott in 1934 by ingenious experiments with

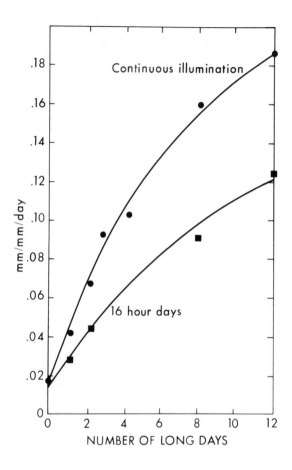

Figure 5. The effect of number of long day exposures, of either 16 h photoperiods (squares) or continuous light (circles), on the rate of inflorescence development in Lolium in subsequent short days (Evans, 1969).

spinach, a long day plant. He used thimbles to impose short day conditions on the apex and light pipes to expose it alone to long days. Flowering occurred only when the leaves were in long days. Subsequent experiments with many short day plants have shown that flowering occurs only when at least some leaf tissue is exposed to short days, for example by enclosing single leaves in envelopes of black paper or aluminum foil. Half-expanded leaves are the most effective ones in Xanthium, but in Perilla and Lolium quite old leaves retain the ability to induce flowering by responding to daylength. The cotyledons alone are sufficient in Chenopodium and Pharbitis, and contribute to floral induction in Anagallis and possibly in Xanthium. Certainly, quite small areas of leaf, less than one to two cm^2 in some cases, are all that is required for photoperiodic induction in several plants.

Although leaves are the primary site for the perception of daylength and shoot tips do not need to be exposed to inductive days, we should not assume that shoot tips are unable to respond directly to changes in daylength. Experiments in India with cultured shoot tips of Cuscuta reflexa, a parasitic plant, indicate that its excised apices can respond to short days. Likewise, Schwabe was able to induce a short day response in Kleinia by covering only the terminal growing points with black plastic caps. Because of the dominance of leaves in the perception of daylength, the consequences of a direct response to daylength by shoot apices have been almost totally neglected and merit further examination.

Darkness or Light, One or Both?

In the very first paper concluding that daylength could control flowering, Tournois

suggested that the length of the night might be
more important than that of the day. Subsequent
work lent weight to what could be taken as a
natural opinion for an ardent young Frenchman.

Garner and Allard construed photoperiodism
as a response to the relative lengths of the
daily light and dark periods, and they carried
out a long series of experiments with alter-
nating light and dark periods of equal duration,
ranging from fifteen seconds to twelve hours.
Long day plants flowered most rapidly under the
shortest cycles, which avoided long dark periods, while short day plants remained vegetative
unless they received long, uninterrupted dark
periods. Or, rather, that is how we can inter-
pret their results in hindsight.

A classical series of experiments in 1938
by Hamner and Bonner clearly established the
central role of the dark period in the photoperiodic processes of <u>Xanthium</u>, from three kinds
of evidence. They showed that the dark period
had to be at least eight and one-half hours long
before flowering could occur, and that this held
even when the overall cycle length was longer or
shorter than twenty-four hours. Plants on cycles
of four hours light/eight hours darkness did not
flower although the light period was much
shorter than the critical daylength of fifteen
and one-half hours. Yet plants in cycles of
sixteen hours light/thirty-two hours darkness
did flower even though the light periods were
longer than the critical length. Thus neither
the length of the light period nor the relative
length of the light and dark periods was as important as the length of the dark period, which
had to exceed a critical duration. We now know
that similar experiments on other short day
plants with a stronger diurnal rhythm would
not have given such a clear cut answer as they
did with <u>Xanthium</u>.

Hamner and Bonner's second line of evidence was that the temperature of the dark period had a much greater effect on flowering response than did that of the light period. Two years later, however, Mann showed that the temperature of the light period could be very important when Xanthium was grown in dull short days, and in 1959 Nitsch and Went discovered that a non-inductive dark period of eight hours at $23°$ C became inductive for Xanthium when the first half of the light period was at a low temperature.

Their third line of evidence has been less eroded by time and has played a crucial role in further analysis of the light reactions in photoperiodism. Exposure to quite low intensity light for only one minute in the middle of an otherwise inductive dark period suppressed flowering completely. Just what Hamner and Bonner had in mind when they tried such "light-break" treatments is not known, but they certainly established Xanthium's requirement for an uninterrupted period of darkness. Brief dark breaks during the day, by contrast, have virtually no effect on flowering.

With long day plants, also, the dark periods seemed to have a predominant effect on flowering, but in this case an inhibitory one. Continuous light gave most rapid flowering, and the longer the dark period and the higher its temperature, the greater was the delay. With plants of henbane, Hyoscyamus niger, one of the most useful long day plants in early experiments, Lang and Melchers found that the higher the temperature, the longer was the daylength required for a flowering response. Once again, however, subsequent experiments with other LDP have shown that day temperature is also important, as is the spectral composition of light during the day.

The question, whether light or darkness

was more important in photoperiodism, was a
natural one to ask in early experiments, and for
a while it seemed to have a clear answer. In
fact, the emphasis on the central role of the
dark period for both short and long day plants
was so strong that it was suggested they should
be referred to as long and short night plants,
respectively. One consequence, which we are
still struggling to shrug off, is that time
measurement came to be thought of as a property
of the dark period alone, in terms of a chain
of chemical reactions, rather than of the interaction between light and dark periods within
each diurnal cycle.

Three

A Transducer for
the Light-Dark Cycles

Action Spectra and Pigment

Hamner and Bonner's finding that flowering in
Xanthium could be prevented by a brief, low-
intensity light break in the middle of a long
dark period provided the key to analysis of the
light reactions controlling photoperiodism. It
also opened a Pandora's box of other sensitive
responses to light.

At the USDA Plant Industry Station at
Beltsville, Maryland, Borthwick, Hendricks, and
Parker constructed, with an old carbon-arc cin-
ema projector and prisms lent by the Smithsonian
Institution, a spectrograph for irradiating
leaves with wavelength bands throughout the
visible region of the spectrum. With such a
setup it was possible to derive an action spectrum for the light break effect, showing the
energy or quanta required at each wavelength to
suppress flowering to a given extent. Such
action spectra do not necessarily indicate the
absorption spectrum of the active pigment because of screening and differential absorption
by other pigments and scattering by cell
walls and organelles. Nevertheless the

first action spectrum, measured in 1945 for the
light break effect in soybeans, showed clearly
that a then unknown pigment was involved, with
strong absorption in the red region, little in
the blue region, and even less in the green
region. A similar action spectrum was then found
for the light break effect in Xanthium. Turning
to long day plants, they found that light breaks
given in the middle of the night to plants grown
in days of sub-threshold length caused flowering
in both barley and Hyoscyamus, with action spectra almost identical with those for the inhibition of flowering in soybean and Xanthium (see
Figure 6, next page).

Thus, the one light reaction seemed to control flowering in both long and short day
plants, and it was subsequently shown by the
Beltsville group to control many other phenomena
as well, such as seed germination, stem and leaf
growth, and pigment development.

Earlier experiments by Flint and McAlister
had shown not only that red light was promotive
of lettuce seed germination but that far red
light--at and beyond the long wavelength end of
the spectrum visible to humans--was inhibitory
to it. An action spectrum for this inhibition
was also determined and found to overlap with
that for promotion. During a seed germination
run on the spectrograph in 1952 it occurred to
the Beltsville investigators to transfer some
Petri dishes of seed from the wavelength band
of maximum action in the red to that in the far
red, and vice versa, so that a saturating exposure in one wavelength was immediately followed
by a comparable exposure in the other. The result was startling. The eventual germination was
determined entirely by the wavelength of the
final exposure: germination if it was red, none
if it was far red. Light was acting like a
switch--red on, far red off--and it was soon

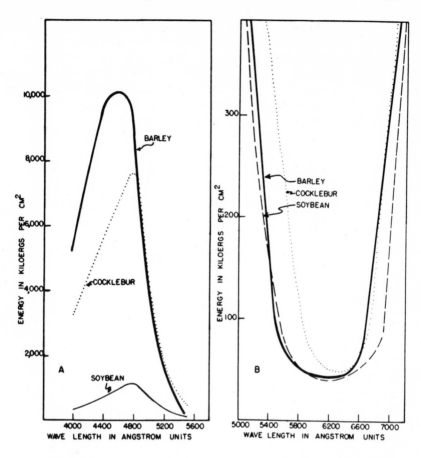

Figure 6. Action spectra for the effect of light breaks during long dark periods on the flowering of two short day plants (soybean and cocklebur) and one long day plant (barley). After Borthwick, Hendricks and Parker (1948).

shown that the germination response could be pushed backwards and forwards repeatedly, with only the final exposure mattering, provided there were no long delays between red and far red exposures.

Naturally enough, experiments were immediately set in hand to see if the light break
effect on the flowering of <u>Xanthium</u> was equally
reversible, and so it proved to be.

To explain these results, Borthwick, Hendricks, and Parker proposed that the receptor
pigment, subsequently called phytochrome, ex-
isted in two forms, one predominantly absorbing
red radiation with a peak at 660 nm (P_r or
P_{660}), the other absorbing far red radiation
with a peak near 720 nm (P_{fr} or P_{720}). Red
radiation would push the reaction

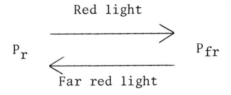

to the right, leaving all the pigment in the
P_{fr} form, while subsequent exposure to far red
light would push all the pigment back to the
P_r form. Because exposure to red light caused
germination and flowering in long day plants
and inhibited the action of an inductive dark
period in short day plants, P_{fr} was assumed to
be the biologically active form of the pigment.

Dark Reversion of Phytochrome

From their experiments on seed germination the
Beltsville workers deduced that P_{fr} did not
necessarily require exposure to far red radiation in order to revert back to the inactive P_r
form of phytochrome. Reversion also appeared to
occur in darkness, at high temperatures.

Thus:

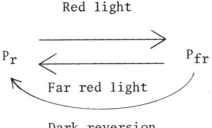

Red light

P_r P_{fr}

Far red light

Dark reversion

With this model derived from the germination experiments in mind, Borthwick and his colleagues then did an experiment destined to have a profound effect on the interpretation of photoperiodic timing for many years. They reasoned that most of the phytochrome would be in the active P_{fr} form at the end of the day and might then undergo dark reversion in the early hours of the night. Perhaps the inductive dark reactions could not begin until nearly all the P_{fr} had reverted to P_r? If so, a brief exposure to far red light at the end of the day, driving all the pigment to the P_r form immediately, should shorten the critical night length of short day plants by the length of time required for dark reversion. In their experiments with Xanthium, Borthwick, Hendricks, and Parker (1952) found the critical dark period to be nine hours when a brief exposure to red light preceded the dark period, but less than seven hours when far red light preceded it.

Thus, the view was developed that in short day plants the inductive dark reactions begin only after the phytochrome P_{fr} has reverted to the P_r form, this reversion being one component of time measurement in the dark period, whereas long day plants require the presence of phytochrome P_{fr} throughout most of the diurnal cycle. This hypothesis retained considerable

explanatory power for many years, and is still
to be found in most text books.

Reversibility provided a ready criterion
for the involvement of phytochrome and could be
explored without access to a spectrograph since
simple plastic or cellophane filters were found
to provide effective sources of red and far red
light. So began something of a scientific gold
rush to extend the reach of the pigment phytochrome, if not our understanding of how it
worked. Occasionally, the inhibitory effects of
red light breaks on flowering were found not to
be reversible, as Nakayama et al (1960) found
with Pharbitis. In this case, however, Fredericq
(1964) was able to show that the failure of
reversibility was because the phytochrome P_{fr}
acted so quickly in the middle of the dark
period that the effects of a red break were
reversible only if far red light was given within thirty seconds of the beginning of the exposure to red light. Escape from reversibility is
usually rather slower than in Pharbitis. In
Xanthium, for example, Downs (1956) found that
the effects of a red light break for two minutes
could be largely reversed by far red provided
not more than fifteen minutes of darkness inter-
vened. There were also cases, in both short day
plants such as Pharbitis and long day plants
such as Lolium, where subsequent exposure to far
red light reinforced rather than reversed the
effect of an earlier red light break. These and
other apparently anomalous features of the
action of phytochrome on the flowering of various plants--of which there is a continuing over-
supply--should not get in the way of the main
line of argument. Many of these anomalies have
been explained as we have come to know more
about the forms of phytochrome and the kinetics
of their transformations.

The pigment was physically detected in

1959, and its chromophore identified in 1966.
Given the seven-year cycle in progress on
phytochrome, evident in the calendar: 1938,
light break effect; 1945, action spectra; 1952,
reversibility; 1959, detection; 1966, identification, I was confident that its role in metabolism would be discovered in 1973, but that did
not prove to be so. Almost all our more direct
knowledge of the kinetics of phytochrome changes
from one form to another comes from work on dark
grown seedling tissues. While this is of value
in understanding photomorphogenesis, it seems to
be of less relevance to the active forms of
phytochrome in the green tissues with which we
are concerned in photoperiodism. Until new methods of assaying active phytochrome in green
leaves have been developed, we must continue to
rely on rather indirect ways of analyzing the
role of phytochrome in photoperiodism.

Phytochrome Transformations

(a) Short day plants. Let us return to the crucial 1952 experiment of the Beltsville group
with Xanthium. Their interpretation of it
created a unified theory for the photoperiodic
reactions in plants but, in doing so, generated
a major paradox. If, as grafting experiments
suggested, the photoperiodic processes of both
long and short day plants resulted in the production of a floral hormone common to both
groups, how could its synthesis require the
absence of phytochrome P_{fr} in short day plants
and its presence in long day plants? Apart from
this paradox, which is not to be ignored, other
more immediate difficulties arose.

The germination experiments had suggested
a rather slow metabolic dark reversion of
phytochrome, almost too slow to fit into a

diurnal cycle, whereas the Xanthium experiments
suggested that it was all over in the first two
hours or so of darkness. As such, it still left
most of the critical dark period length to be
accounted for.

Moreover, although brief exposure to far
red light appeared to shorten the ensuing critical dark period in a few other short day plants,
the effect was often slight and very dependent
on the length of the preceding light period. In
fact, far red light at the end of short light
periods was more often inhibitory to flowering
than promotive, as in Xanthium, Pharbitis, Kalanchoë, Chenopodium rubrum, and Lemna perpusilla. This inhibition may be pronounced even
when the following dark period is very long.

With longer light periods and shorter dark
periods, far red light at the beginning of the
dark period occasionally promotes flowering, as
in Xanthium, but is often inhibitory. In fact,
far red exposures may be inhibitory not only at
the beginning of the dark period in Pharbitis,
but for several hours later. This phenomenon is
also evident in Chenopodium and even in Xanthium
when it is given only short periods of daylight.
It is an important observation because it im-
plies not only that phytochrome P_{fr} is needed
in these short day plants throughout the early
hours of darkness but also that it is still
present then. If so, metabolic dark reversion to
the P_r form has not occurred as we should expect
from the Beltsville theory.

Clearly, we need better evidence of whether
and when dark reversion of phytochrome occurs
in the green leaves of short day plants. There
have been many experiments on how the effect of
a red light break changes in the course of a
night, but they give us no more than a rough
indication of when phytochrome reverted to the
P_r form. A technique developed by Cumming,

Hendricks, and Borthwick (1965) is far more useful. Instead of merely giving groups of plants a brief exposure (say one to three minutes) to red light at various times during the dark period, or to far red light, additional groups are exposed to various mixtures of red and far red light. The dark period length is adjusted, where possible, to give an intermediate flowering response in the unexposed control group. Of the various mixtures of red and far red light given at any one time there will be one which scarcely changes the relative proportions of P_{fr} and P_r present in the leaf just before the light break is given, and this should have no effect on the flowering response, whereas the others will promote or inhibit depending on the requirement for phytochrome P_{fr} at that time. The percentage of red light in the red plus far red mixture which gives a null reaction (i.e., no effect on flowering) can then be plotted for various times during the dark period to indicate when phytochrome reversion occurs. Of course, the exact relation between percentage of red light and percentage of phytochrome present as P_{fr} cannot be determined, but a fall in the percent of red light giving a null reaction should indicate a corresponding fall in the percent of phytochrome in the P_{fr} form and, therefore, the timing of dark reversion.

This technique has been used with two SDP, Pharbitis and Chenopodium, and with one LDP, Lolium, and some of the results obtained with them are summarized in Figure 7 (next page). The left side of the figure shows the shift in the percent of red light giving a null reaction for light breaks at various times after the beginning of darkness. With Pharbitis this remained high for five hours, but then fell rapidly, indicating dark reversion of phytochrome to the P_r form at that time. In

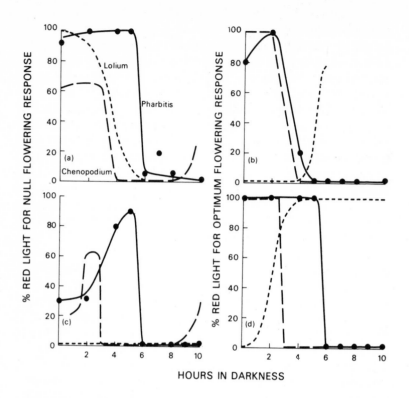

Figure 7. Changes during darkness in:
(a,c) the percentage of red light $\left[\text{i.e., } \dfrac{R \times 100}{R+FR}\right]$
in light breaks having no effect
on flowering.
(b,d) the percentage of red light in light
breaks having an optimum effect on flowering.
(a,b) the upper half of the figure refers to
plants briefly exposed to red light at the
beginning of darkness.
(c,d) the lower half refers to plants exposed
to light rich in far red at the onset of
darkness.
    From data of Evans and King (1969) for
Pharbitis, King and Cumming (1972b) for Chenopodium, and Evans (unpublished) for Lolium.

Chenopodium dark reversion occurred somewhat
earlier, after three to four hours of darkness.
Clearly, reversion does not begin with the onset
of darkness, and in Pharbitis it occurs towards
the middle of the critical dark period. Moreover, Figure 7b indicates that not only is most
of the phytochrome present in the P_{fr} form
during the early hours of darkness, but also
that this is required for optimum flowering in
both short day plants, a very different situa-
tion from that envisaged in the Beltsville
model.

One of the most striking features of these
experiments is shown in Figure 7c. If plants were
exposed briefly to light rich in far red at the
beginning of darkness, to drive most of the
phytochrome to the P_r form, there was a progres-
sive rise in the percent of red light giving a
null reaction up to the time of dark reversion,
i.e., the proportion of phytochrome in the P_{fr}
form increased in darkness, the opposite of what
we expect from dark reversion. When we discuss
endogenous rhythms later, we shall see that
there is evidence that this "inverse reversion"
can recur, with a circadian rhythm, in long dark
periods.

These results can explain many of the apparent anomalies accumulated in earlier flowering experiments, but these need not concern us
here. What should be stressed, however, is that
the light-off signal cannot be the only one controlling the transformations of phytochrome in
darkness, as we shall discuss later.

(b) Long day plants. The Beltsville scheme envisaged LDP as requiring most of the phytochrome
to remain in the P_{fr} form throughout the day for
maximum flowering to occur. Two complications
soon became apparent. The first, of which the
Beltsville workers were well aware, is that red

light is not very effective for extending the
daylength for a great many long day plants, compared with light with approximately equal proportions of red and far red radiation. This is
experimentally convenient in that light from incandescent lamps is usually very effective as an
extension of the day to induce flowering in LDP,
but it has yet to be fully explained. The optimum proportion of red light varies somewhat between LDP, being much higher in Hyoscyamus than
in sugar beet, with Lolium in between. It also
varies with daylength and temperature, tending
to be lower the longer the day and the higher
the temperature. Action spectra for flowering in
long day plants therefore depend on the experimental setup. For brief light breaks in the middle of a marginally inhibitory dark period, red
light with a wavelength of 660 nm is most effective, as in the classical Beltsville action
spectra. But for prolonged extensions of the
photoperiod, peak action may be found at wavelengths between 695 and 720 nm.

Consider the in vitro action spectra for
the P_r and P_{fr} forms of phytochrome when ex-
tracted, purified, and free of screening by
other pigments like chlorophyll (see Figure 8,
next page). Red light of 660 nm wavelength
drives the interconversion mostly to the P_{fr}
form, but even at that wavelength about one
fifth of the pigment will remain in the P_r form
because absorption by the P_{fr} tail will drive
the reaction partly in the opposite direction.
At wavelengths around 690 nm the two forms of
the pigment will be present in approximately
equal amounts. Moreover, there will be continual
cycling between the P_r and P_{fr} forms, so that
the many short-lived intermediate forms now
known to occur between them will also be present.
At wavelengths beyond 720 nm virtually all the
phytochrome is present in the P_r form.

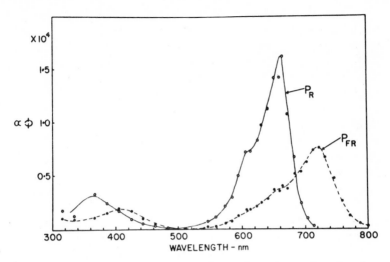

Figure 8. Action spectra for the photochemical transformations of phytochrome P_r and P_{fr} in vitro. α is the extinction coefficient, θ the quantum yield. (Butler et al. 1964)

Many explanations of the requirement by LDP for an approximately equal mixture of red and far red light are therefore possible. LDP may need P_{fr} to be present throughout the day but at less than the level of full conversion. Or perhaps the short-lived intermediate forms of phytochrome play a role. Or perhaps LDP, like SDP, have a requirement for both high-P_{fr} and low-P_{fr} processes, which in the case of LDP can both occur at some intermediate level of P_{fr}.

This brings us to the second problem with the Beltsville scheme for LDP. Since they can flower rapidly in continuous incandescent light, LDP clearly do not require a light-dark cycle or an alternation in P_{fr} level, but where this does occur there is much evidence that far red light and a low P_{fr} level is favorable about

eight to twelve hours after the beginning of
high-intensity light. Many experiments with
Hyoscyamus at Wageningen, by Stolwijk and
Zeevaart, de Lint, and El Hattab, have demonstrated a promotion of flowering by far red
light at the end of a short day. In fact,
Hyoscyamus will even flower in a daylength less
than the critical when the day is terminated by
a brief exposure to far red light. In Lolium,
too, Vince (1965) showed that whereas extension
of a short day with eight hours of red light was
ineffective, extension with seven hours of far
red followed by one hour of red was highly
effective. The Lolium results in Figure 7 indicate why; they show that a low proportion of
phytochrome in the P_{fr} form is optimal for
flowering throughout the early hours of darkness after a short day, after which a high
proportion of P_{fr} is needed.

Working Hypothesis

These experimental results lead us back to a
more unified view of the photoperiodic reactions, preserving the symmetry of the Beltsville
hypothesis without its built-in paradox of how
opposite paths lead to the same end. Flowering
in both short and long day plants is dependent
on a process or processes favored by low P_{fr}
levels and also on a high P_{fr} process. The low
P_{fr} process is particularly limiting in Xanthium
and some other SDP, the high P_{fr} process more so
in LDP. But in SDP like Pharbitis and
Chenopodium, and in LDP like Lolium and
Hyoscyamus, limitations by both processes are
readily exposed.
       What, then, is the essential difference be-
tween SDP and LDP? It seems to lie in the optimal timing for the two processes in relation to

the high-intensity light period. In _Pharbitis_
and _Chenopodium_ the high P_{fr} process is opti-
mally consummated in darkness immediately after
the end of the light period (Figure 7), the low
P_{fr} process occurring later. On the other hand,
in the LDP _Lolium_ and _Hyoscyamus_, the low P_{fr}
process should succeed the light period, while
the high P_{fr} process is required later, as
Figure 7 b,d indicates.

Whether these are actually processes rather
than alternate phases of an endogenous rhythm
will now be considered.

Four

Endogenous Rhythms and
Other Light Reactions

Circadian Rhythms

Rhythms in the inclination of leaves and petals
which persist even in constant conditions (hence
endogenous), with a periodicity of about one day
(hence circadian), have long been recognized in
plants. As early as 1936 Bünning suggested that
time measurement in photoperiodic reactions
might be associated with endogenous rhythms,
and reassertion of this idea at various times
over the next twenty years still led to remark-
ably little interest in it. For one thing, the
physiological basis of such endogenous rhythms
was, and remains, obscure so that there was
little by way of explanation to be gained from
the association. Also, by the 1950's most hopes
for the elucidation of time measurement were
being pinned on phytochrome reversion and related phenomena, on hourglass rather than rhythm
methods of control.

One problem for hourglass models based on
the time required for a sequence of reactions to

occur is posed by the rather slight effect of temperature on the length of the critical dark period in Xanthium (see Figure 9). Such a low temperature coefficient would be more representative of physical processes than of a series of chemical reactions. However, it is possible to devise systems of chemical reactions which are temperature-compensated, particularly if both promotive and inhibitory processes are involved. Moreover, some photoperiodically sensitive plants, such as Pharbitis, display a far greater degree of temperature dependence in the length of their critical dark period than Xanthium does (see Figure 9, next page). In those plants whose critical daylength is little affected by temperature, flower induction may be dominated by rhythmic processes, whereas "hourglass" processes may predominate in those with a marked temperature dependence.

A further problem with phytochrome reversion as a timer arises since we now are aware that reversion may not begin until after several hours of darkness. What is it that measures the initial time interval?

In considering the possible role of endogenous rhythms in photoperiodism, the first thing to be said is that flower induction in both LDP and SDP can occur in the absence of any environmental rhythm. This is well known for many LDP, and has been found with several SDP grown in continuous light of a spectral composition which drives most of the phytochrome to the P_r form. Examples include Chenopodium rubrum, strawberry and Lemna perpusilla. At low temperatures quite a number of short day plants, e.g., Pharbitis, will flower even in continuous light with a high proportion of red radiation.

Thus, a rhythmic environment is not a prerequisite for flower induction in photoperiodically sensitive species, but it may certainly

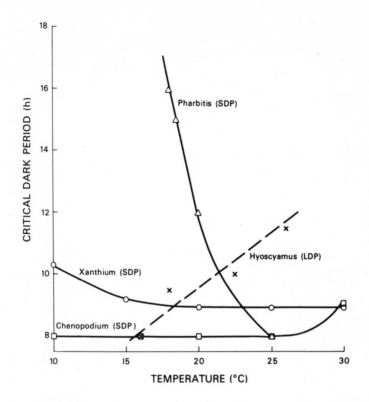

Figure 9. Effect of temperature on the length of the critical dark period for flowering in SDP Xanthium (data of Salisbury, 1963), Pharbitis (data of Takimoto and Hamner, 1964), and Chenopodium (unpublished data of R. W. King), and in LDP Hyoscyamus (data of Lang and Melchers, 1943).

entrain the inductive processes and amplify the flowering response determined by hourglass mechanisms. For example, in a long series of experiments with Pharbitis, Takimoto and Hamner (1964)

found that when plants were transferred from
continuous light to darkness at suboptimal temperatures, flowering increased progressively
with increase in the length of the dark period,
presumably because the hourglass reaction was
limiting. Previous exposure to an entraining but
non-inductive light-dark cycle superimposed a
rhythm on this response.

Rhythmic Flowering Responses

Circadian rhythmic effects on flowering have
been established in several ways. Groups of
plants may be exposed to one or more cycles com-
prised of a fixed light period combined with
dark periods ranging in length from a few to
eighty or more hours. Survival of the plants
(and of the experimenter) through such long dark
periods is the main limitation. Figure 10 illustrates the results of such an experiment with
Chenopodium. Comparable results have been obtained with other SDP, such as soybeans, and
with the LDP Hyoscyamus.
    Alternatively, all the plants may be ex-
posed to a given long dark period, which is
interrupted by a brief light break at different
times in different treatment groups. Many such
experiments have been done with short day plants
such as Chenopodium, soybean, Pharbitis, and
Kalanchoë, but fewer with long day plants such
as Hyoscyamus, Lemna gibba, and Anagallis. The
results of an experiment with Chenopodium are
also given in Figure 10 (next page).
    Results such as these provide clear evi-
dence that the flowering process, like many
others in plants, is influenced in its expression by endogenous rhythms. Such rhythms imply
the existence of a circadian clock, and it has
been suggested that the readily observed overt

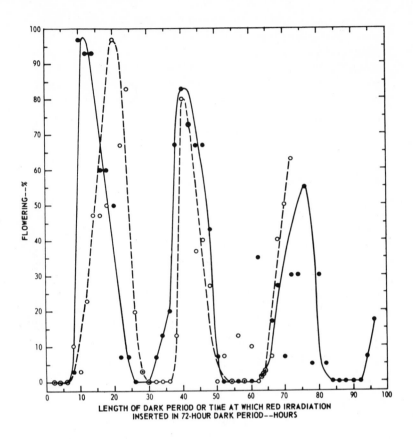

Figure 10. Flowering response of Chenopodium rubrum following exposure either to a single dark period of varied length (solid line and symbols) or to a 72-hour dark period interrupted at various times by 4 minutes of red light (Cumming et al. 1965).

rhythms in leaf and petal movements or CO_2 output may be used as clock hands for the rhythm in inductive processes which only becomes apparent several days later when flowers are formed. However, there may not be a fixed phase relation

between the overt rhythm and that in the induction of flowering. Indeed, Wagner and Cumming
(1970) have shown that at least one circadian
rhythm, in accumulation of the pigment betacyanin, displays a phasing which does not always
correlate with the flowering rhythm. Hillman
(1975) argues that the overt rhythm need not
necessarily correlate with the inductive rhythm,
since there may be several rhythms and since
flower induction may occur when two or more of
these are brought into effective relationship
with one another by environmental entrainment.

To my mind the most important question con-
cerning photoperiodism and endogenous rhythms,
whether such rhythms constitute the time-
measuring photoperiodic processes in leaves, has
yet to be answered. On present evidence it is
just as feasible that the rhythmic flowering
responses are due to a rhythmic responsiveness
of the shoot apex to the inductive stimulus.
Consider the sequence of events in a short day
plant such as Pharbitis when it is exposed to a
long dark period. After about twelve hours of
darkness it will have reached its critical night
length. As we shall see later (p. 60), the induced leaves then begin to export floral stimulus, and by the eighteenth hour sufficient
stimulus to induce a full flowering response
will have been transported from the leaves even
when they remain in darkness. Thus, the rhythmic
inhibitory effects of still longer dark periods
can hardly be due to rhythmic effects on the
production of the photoperiodic stimulus. In
Xanthium, Searle showed that export of the
photoperiodic stimulus from the leaves is un-
affected by light or darkness and takes place
rapidly after twenty hours in darkness. In this
case also, the rhythmic inhibitory effects of
still longer dark periods are likely to be an
expression of rhythmic changes in apical

response, possibly associated with a rhythmic mobilization of reserves. Searle's experiments with Xanthium, unlike those of Chorney et al. (1970), showed no rhythmic flowering response to the length of the dark period, presumably because the leaves below the darkened leaf in his experiments were kept in continuous light, yielding a continuous supply of assimilate to the shoot apex.

Likewise, King and Cumming (1972 a) concluded that the rhythmic oscillation in flowering of Chenopodium is a response to the time at which light terminates darkness, which suggests that the apex is more likely than the leaf to be the responsive organ.

However, the light-off signal at the beginning of the dark period may also initiate many rhythmic responses. One of these, almost certainly in the leaves, involves the transformations of phytochrome. We have already referred to the experiments of King and Cumming (1972 b) which showed that virtually all the phytochrome in Chenopodium cotyledons reverted to the P_r form after about four hours in darkness. Yet after eight hours of darkness the percent of red light giving a null reaction to light breaks began to rise again, indicating a reappearance of P_{fr} phytochrome. In the experiments of Cumming et al., (1965) the percent red giving a null reaction was high after eighteen hours of darkness, then fell to the thirtieth hour, rose and fell again by the fifty ninth hour of darkness, after which it rose again. These results strongly suggest a rhythmic reversion and reappearance of phytochrome P_{fr} even in the absence of the diurnal light-dark cycles which normally drive the phytochrome transformations. If endogenous rhythms control the timing of phytochrome transformations in leaves, they may well play an important role in the time-measuring reactions

which specify the length of the critical dark
period. But the evidence is still no more than
suggestive.

Other Light Reactions

The discussion above has concentrated on phytochrome transformations as the controlling light
reactions in photoperiodism. From time to time
as complications have arisen in the interpretation of responses to light treatments in terms
of phytochrome, other pigments and light reactions have been invoked, particularly to explain
flowering behavior in long day plants. By and
large, however, many of these apparent anomalies
have proved to be explainable in terms of phytochrome as we have come to understand more of
the complexities of its action. This is particularly true of the many responses to far red
light which did not seem to fit the Beltsville
scheme. Many responses to blue light are also
explainable now that we know that phytochrome
interconversions are driven in both directions
by most wavelengths in the blue region (see
Figure 8, page 43). For many long day plants,
therefore, blue light can be as effective as
light of 700 nm wavelength.
      In some long day plants, particularly
Cruciferae like Sinapis alba, blue light is far
more effective for the induction of flowering
than either red or far red light whether given
as photoperiod extensions or as night breaks.
Funke and several Dutch investigators had shown
this early on, and it has been confirmed in more
recent experiments with light sources of higher
resolution and purity. It is possible to explain
such responses to blue light in terms of phytochrome, at least in dark grown seedlings, but an
alternative is that a separate blue-far red

high-energy reaction of photomorphogenesis is
involved.

The products of photosynthesis are likely
to be substrates for photoperiodic reactions in
the leaves and at the shoot apex. It is not sur-
prising, therefore, that they may sometimes limit
the flowering response, as indicated by the inhibitory effects of exclusion of CO_2 or application of DCMU (3(3, 4 dichlorophenyl)-1,1 dimethyl
urea), a specific inhibitor of photosynthesis.
For example, many early experiments with short
day plants such as Xanthium and soybean showed
that at least a brief period in high-intensity
light in the presence of CO_2 is required for a
subsequent inductive dark period to be maximally
effective. Another such period is also required
by Xanthium at the end of the dark period to
assist the translocation of the floral stimulus
and the response of the shoot apex to it. For
several years these requirements were unneces-
sarily sanctified as the first and second high-
intensity light processes in Xanthium, but
photosynthesis by any other name would suffice.

In recent experiments with Sinapis alba, a
long day plant, Kinet et al. (1973) found exclusion of CO_2 during the first eight hours of light
in a long day to inhibit flowering, whereas exclusion of CO_2 later in the day did not do so.
Thus, although the long day flowering response
was dependent on photosynthesis for its expression, the long day photoperiodic reaction did
not require continuing photosynthesis.

Five

Transmissible Stimuli and Inhibitors of Flowering

Translocation of the Photoperiodic Stimulus

Working with spinach plants, Knott (1934) showed that although it is the shoot apex that ultimately responds by initiating flower primordia, it is the leaves that perceive the daylength. He concluded that "the part played by the foliage of spinach in hastening the response to a photoperiod favorable to reproductive growth may be in the production of some substance, or stimulus, that is transported to the growing point." Comparable experiments by Chailahjan, Moshkov, and Psarev in Russia led Chailahjan to advocate, in 1936, the existence of a floral hormone. Grafting experiments, which we discuss later, provided more direct evidence supporting the concept of a mobile floral stimulus, but gave little information about the nature and speed of its movement.

Although some early reports suggested that the floral stimulus could pass through water or across tissue paper, more careful experimentation indicated that its movement was probably

confined to living tissue. By severing the
phloem connections between leaves and growing
point in the SDP Perilla, Chrysanthemum, and
Xanthium, first Chailahjan and then the Withrows
succeeded in preventing movement of the photoperiodic stimulus. Other observations also indicated that the stimulus moved in the phloem. For
example, in plants with vascular systems having
only limited cross connections, exposure of only
one leaf to long or short days can lead to the
lop-sided induction of flowers, reflecting the
vascular pattern, as Harder found with Kalanchoë. Also, in experiments with two-branched
plants, one in inductive daylengths and the
other not, flowering could be induced in the
receptor branch only when its leaves were removed. To explain such a response, it was assumed that the photoperiodic stimulus moved in
the phloem, along with the sugars. Receptor
shoots with all their leaves intact would not
import sugars from other shoots (unless kept in
darkness) thereby missing out on the photoperiodic stimulus which was assumed to move by mass
flow with the sugars.

Such results led to the early realization
that in grafting experiments the receptor scions
should be defoliated to encourage transmission
of the floral stimulus from the stock. This requirement was not universal, however. The stimulus generated by short days in Maryland Mammoth
tobacco plants was found by Zeevaart (1958) to
move only into defoliated scions, whereas that
generated by long days in Nicotiana sylvestris
can move either up or down into leafy scions.

Timing and Speed of Movement

As flowering experiments came to be concentrated
more on plants requiring only one inductive

cycle, there developed a need to pinpoint more closely the time at which translocation of the photoperiodic stimulus was taking place. Three groups working with <u>Xanthium</u> hit upon the same technique at about the same time in 1954.

Plants with a single leaf were exposed to an inductive dark period, and that one leaf was then cut off at various times thereafter in separate groups of plants. Doing this immediately after the dark period prevented the plants from flowering, but the longer the leaf remained attached, the more stimulus was exported before cutting and the greater was the flowering response, as can be seen in Figure 11. Enough

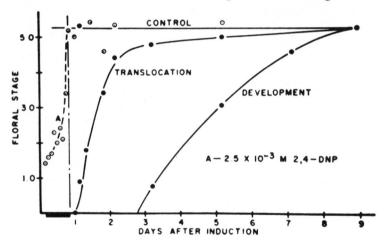

Figure 11. Flowering response in <u>Xanthium</u> as influenced by (1) the time at which the leaf exposed to the one short day was cut off, putting an end to its export of the floral stimulus (2) the time at which buds were dissected, indicating the rate of progress through the floral stages illustrated in Figure 4 (3) the time at which 2.5×10^{-3}M 2,4-dinitrophenol, which uncouples respiration, was applied to the leaves (Salisbury, 1963).

stimulus to support a maximal rate of flower
development was usually exported by the end of
the day following the dark period. However, as
noted earlier, even Xanthium and Pharbitis
leaves kept in darkness can export the floral
stimulus over the same interval as leaves in
light.

Such experiments tell us *when* the stimulus
is exported but not how fast it moves. If we
could ascertain its speed of movement we could
estimate when it should reach the growing point
and thus identify the time at which to look for
the earliest consequences of its arrival there.
To gain this additional information we need only
to elaborate the technique described above. Instead of severing all the leaves at one point,
say the base of the blade, we need another set
of plants in which the translocation pathway is
severed at a fixed distance further along it.
In this way, the time taken for the amount of
stimulus causing a certain flowering response
to traverse a known distance can be ascertained
in spite of our continuing ignorance of the
identity of the stimulus. Such an experiment was
first done with Lolium, with the results shown
in Figure 12 (next page). In this case the base
of the one leaf blade exposed to the long day
was wrapped in foil and severed at various times
in one of three positions. The speed of translocation of the long day stimulus in several
such experiments varied from 1-2.4 cm/hr. We
also made concurrent measurements of the speed
of translocation of ^{14}C-labeled photosynthate
over the same pathway and found it to vary from
77-105 cm/hour. Thus, the long day stimulus
could not be moving by mass flow with the sugars
in Lolium. This conclusion was supported by the
finding that Lolium leaves which were only one
quarter expanded were effective exporters of the
long day stimulus to flowering, yet exported

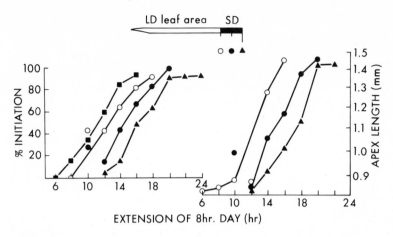

Figure 12. Translocation of the long-day stimulus in Lolium: the effect of time of removal of plants to darkness (■) or of cutting off the only leaf blade exposed to the one long day at its base (▲) or 4 (●) or 8 (○) cm above the base, on the flowering response in terms of either % of plants initiating inflorescences or apex length at dissection (Evans and Wardlaw, 1966).

hardly any ^{14}C-labeled assimilate to the shoot apex.

    In view of these rather unexpected findings we then measured the speed of translocation of both sugars and short day stimulus down the stem of Pharbitis. The experimental setup differed somewhat from that for Lolium, but temperature and light conditions were comparable. In this case, however, the daylength stimulus to flowering moved at an average speed of 24-37 cm per hour over several experiments, and simultaneous ^{14}C-assimilate movement was at 33-37 cm per hour. Thus, the short day stimulus to flowering in Pharbitis was being translocated more than ten times faster than the long day stimulus in

Lolium and very likely in association with the
photosynthetic sugars in the phloem, a conclusion supported by the finding that leaf size in
Pharbitis had similar effects on the export of
both sugars and floral stimulus.

These experiments are in urgent need of extension to other one-cycle plants, as two quite
different interpretations of them are possible.
One is the bane of flowering physiology, that
different plants differ in different ways. The
other is that the photoperiodic stimulus to
flowering in LDP differs from that in SDP. Many
grafting experiments appear to refute the second
alternative, but other explanations of them are
possible. It is a pity, therefore, that so few
plants appear to be suitable for determinations
of the speed of translocation of the photoperiodic stimulus.

Are There Transmissible Inhibitors of Flowering, as Well as Stimuli?

Evidence for the transmission of a stimulus to
flowering from photoperiodically induced leaves
has now been obtained for many plants, although
there remain some in which it has been sought
but not found in either grafting or defoliation
experiments. Is it possible, as Gregory first
wondered in 1936, that flowering in these plants
is actively inhibited by non-inductive daylengths rather than stimulated by inductive
ones? Or perhaps there are plants in which
photoperiodic inhibitors and stimuli interact
at the shoot apex to determine the flowering
response.

(a) Short day plants. In Kalanchoë Harder showed
quite early that leaves in long days are inhibitory to flowering, but only when they are

inserted between the leaves exposed to short
days and the responding meristem. Moreover, they
are most inhibitory when in the same orthostichy
as the short day leaves, i.e., when connected to
the same vascular traces up the stem. Various
interpretations of Harder's observations are
possible. The long day leaves, being closer to
the growing point (since only then are they inhibitory), may supply it with the requisite assimilates so that there is little assimilate
flow to the growing point from lower leaves in
short days. Studies of the distribution of ^{14}C-
labeled assimilates in many plants have yielded
patterns of behavior like this, particularly for
leaves in the same orthostichy. If the short day
stimulus moves only in association with assimilates, the inhibitory effects of such long day
leaves are readily explained without postulating
a transmissible inhibitor for them.

Alternatively, the upper leaves joined to
the same vascular traces might act as a sink for
the short day stimulus. However, such upper leaves
are unlikely to import assimilates from leaves
below them unless they are still expanding or
are kept under low intensity light. The presence
of young expanding leaves in non-inductive daylengths near buds is often notably inhibitory
to flowering in short day plants, such as
Xanthium. They could well be acting as sinks for
the floral stimulus, but they may also have an
adverse effect through their export of auxins,
gibberellins, and other substances to the nearby shoot apex.

In some recent experiments with Kalanchoë,
Schwabe (1972) found that sap expressed from
leaves in long days, but not sap from short day
leaves, was inhibitory to flowering. Pryce
(1972) has gone on to identify an inhibitor of
flowering in Kalanchoë as gallic acid. This
appears to be present also in the leaves of

Kalanchoë in short days, but in an inactive,
non-dialyzable form. It also inhibits flower
development in the LDP Viscaria candida. Pryce
suggests that gallic acid may bind and inacti-
vate the floral stimulus, but there is no evidence for this nor for export of gallic acid from
long day leaves.

The nature of long day inhibition in Xanthium has been recently reexamined by Gibby and
Salisbury (1971) in an ingenious series of experiments analyzing why long day conditions
given to the basal half of a leaf so effectively
prevent floral induction by the distal half in
short days. The requirements for an inhibitory
long day proved to be an exact complement of
those for an inductive short day. For example,
very low light intensities were sufficient for
both effective long day inhibition and suppression of the inductive dark reactions. Their main
conclusion, for our present purposes, was that
the long day inhibition in Xanthium was likely
to be localized within the leaf itself, and not
transmissible.

Likewise, experiments on Perilla by King
and Zeevaart (1973), extending earlier ones by
Chailahjan and Butenko in Russia, did not produce any evidence of a transmissible inhibitor
of flowering from long day leaves. All the inhibitory effects of long day leaves on flowering
in Perilla could be explained in terms of their
effects on the pattern of distribution of assimilates and of the short day stimulus moving
with them. Consequently, long day leaves were
inhibitory only when they were exposed to high
light intensities. Within one day of their removal, sufficient stimulus to induce a maximal
flowering response could be exported from the
induced leaf, there being no inhibitory after-
effect of the long day leaves.

Although recent work with these three SDP

yields no evidence of a transmissible long day
inhibitor of flowering, earlier experiments by
Guttridge with strawberries do so. Flowering in
short days of daughter plants which were attached to mother plants under conditions
favoring translocation to the daughter, was inhibited when the mother plant was exposed to
long days or light breaks. By contrast, flowering in the mother plants in short days could
be accelerated by defoliation of connected
daughter plants in long days. These and many
other effects suggest that flowering in strawberries is controlled by the presence or absence of a transmissible long day inhibitor,
many of whose effects are closely mimicked by
gibberellic acid.

Experiments with another SDP, the tropical
grass Rottboellia exaltata, also provide clear
evidence of a transmissible long day inhibitor
of flowering. In Rottboellia the flowering response to exposure of the uppermost leaf to six
short days could be greatly reduced by exposing
the leaves *below* it to only one long day towards
the end of short day induction. These results
are of particular interest because long days·
given to the lower leaves near the start of
short day induction increased the flowering response. Thus, in the course of induction the
interplay between the short day stimulus and
the substance exported by long day leaves
changed from synergistic to antagonistic.

Experiments by Fratianne with soybeans
linked by parasitic dodder also provide evidence
of a transmissible long day inhibitor of flowering. Plants in short days linked to others in
long days flowered later than plants not so connected. On intact hosts the dodder itself flowered only in short days, but when the host
soybeans were defoliated they flowered also in
long days.

(b) Long day plants. Among LDP, three examples may be considered. Plants of Hyoscyamus kept in short days never flower as long as they remain intact, but do so rapidly once their leaves are removed. Such a response could be due to flower induction being under the control of an inhibitor exported by leaves in short days. Lang, who did the experiment, prefers the alternative explanation that the leaves diverted, or acted as a sink for, the floral stimulus, but this explanation can hardly be used for experiments by Withrow and his colleagues with spinach. In these, lower leaves in short days inhibited the flowering response to upper leaves in long days. I therefore tried a similar arrangement in experiments with Lolium. The uppermost leaf was exposed to one long day, while the lower leaves remained in short day conditions. The results are shown in Figure 13 (next page). The greater the area of leaves in short days, the more the flowering response to the long day leaf was reduced. Since the short day leaves were below the one exposed to the long day, they could hardly interfere with translocation from it to the shoot apex, a point subsequently confirmed by analysis of the distribution of ^{14}C-labeled assimilates. Thus, the short day leaves presumably exported a substance inhibitory to flowering. The results given in Figure 13 indicate that this export proceeded throughout the subsequent day. Abscisic acid, which inhibits flowering in Lolium, was considered a candidate for the putative short day inhibitor, but this now seems unlikely.

An important implication of the experiments with spinach and Lolium is that the short day inhibitor of flowering, like the long day stimulus, can move independently of assimilate flow since it can reach the shoot apex from lower leaves in the presence of upper leaves. Thus, it

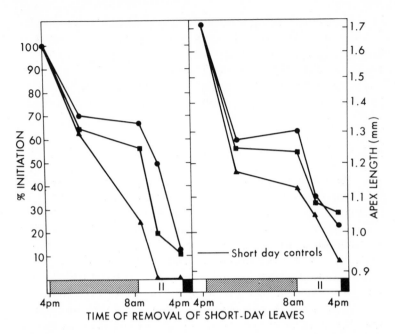

Figure 13. Translocation of the short day inhibitor in <u>Lolium</u>. Plants were induced to flower
by exposure of the uppermost leaf to one long
day. The lower leaves, increasing in area from
● through ■ to ▲, remained in short days and
were cut off at various times during and after
long day induction (Evans, 1960).

is unlikely to be the same substance as the
short day stimulus in SDP, which would have
been an intriguing possibility. But it is clear
that transmissible inhibitors of flowering can
be produced by leaves in non-inductive daylengths in at least some SDP and LDP. In some,
like the strawberry, such inhibitors may have
a dominant influence on floral induction. In
others, like <u>Lolium</u>, they interact in a quantitative way with a photoperiodic stimulus to
flowering. In yet others, like <u>Xanthium</u> and

<u>Perilla</u>, only a photoperiodic stimulus has been
detected. However, it has yet to be shown that
the transmissible inhibitor apparently produced
in non-inductive daylengths is not also made in
inductive daylengths, i.e., that its formation
is photoperiodically controlled.

Graft Transmission of the Stimulus to Flowering

Grafting experiments have had a powerful influence on our interpretations of how daylength
controls flowering. The earliest of them provided strong support for the concept of a flower
hormone and later ones for the identity of that
hormone in different photoperiodic response
groups and genera of plants. Zeevaart (1962)
used the technique to show that juvenile insen-
sitivity in the photoperiodic induction of
<u>Bryophyllum</u> was due to lack of responsiveness
of the early leaves and not of the young shoot
apex, which could soon be brought to flowering
by grafting to an older plant. I had previously
deduced this for <u>Lolium</u>, but could not prove it,
whereas Zeevaart's elegant grafting experiments
soon settled the matter. However, many grafts
fail to demonstrate transmission of the flowering stimulus, even in the hands of experienced
practitioners. From a long series of interspe-
cific grafts with <u>Silene</u>, for example, van de
Pol (1972) concluded that successful transmission was the exception rather than the rule.
Vascular restrictions on assimilate distribution, discussed in the previous section, and
adverse effects of high temperatures or light
conditions may all conspire to cause failure of
transmission. Therefore, it is the successful
grafts on which we should concentrate.
     The first successful graft transmissions of
a photoperiodically-controlled flower stimulus

were made in 1936 by Kuijper and Wiersum in
Holland with soybeans and by Chailahjan and
Yarkovaya in Russia with Perilla. In these experiments, as in many that followed, a leafy
stock of a photoperiodically sensitive plant in
inductive conditions could, through graft contact, cause flowering in a receptor scion in
non-inductive daylengths, particularly if the
receptor was defoliated. Different varieties,
species, and even genera of plants with the
same photoperiodic response have been shown to
transmit the floral stimulus to one another in
experiments of this kind. Van de Pol (1972),
for example, used such experiments to examine
differences between varieties that flowered
early and those that flowered later. For Xanthium, ten days of grafting contact were sufficient to induce full flowering in early scions,
whereas more than twenty days of contact were
needed by late scions, using medium donors in
all cases. Comparable results were obtained
with early and late flowering varieties of
Kalanchoë, suggesting that the early flowering
varieties required less floral stimulus to induce a full flowering response in them.

Chailahjan soon proceeded beyond the
grafting of induced and vegetative plants of
the same daylength response group, and he induced flowering in the short day plant Helian-
thus tuberosus in long days by grafting to day-
neutral H. annuus donors. Moshkov went even
further and induced flowering in SDP Maryland
Mammoth tobacco in long days by grafting it to
the long day plant Nicotiana sylvestris. In the
intervening years many grafts which successfully
induced flowering in plants of one response type
from photoperiodically induced donors of a
different type have been reported and have been
comprehensively reviewed by Lang (1965). Among
recent examples van de Pol (1972) induced

flowering in SDP <u>Kalanchoë</u> with LSDP <u>Bryophyllum</u>
as donor, in LDP <u>Silene</u> with SDP <u>Perilla</u> as
donor, and in SDP <u>Xanthium</u> with LDP <u>Rudbeckia</u> as
donor. This latter graft was not very effective,
and Jacques (1973) also found LDP <u>Blitum</u> to be
less effective as donors to SDP <u>Chenopodium</u> receptors than the reverse arrangement, leading
her to wonder whether there was a lower level of
floral stimulus in long day plants. An explanation of this kind could account for some past
failures to induce flowering in grafts. For example, Zeevaart (1958) found that by using a
day-neutral tobacco variety as donor, flowering
could be induced in LDP <u>Nicotiana</u> <u>sylvestris</u> in
short days, but not in SDP Maryland Mammoth
tobacco in long days.

The major conclusion drawn from these experiments was that there was one floral stimulus, which Chailahjan called florigen, common
to all plants of all photoperiodic classes. In
the enthusiasm for making new graft combinations, the requisite controls have not always
been made, but in four cases (see Lang, 1965)
it has been shown that flowering in the receptor
occurred only when the donor was in inductive
conditions.

Few of these control grafts have been made,
since they are awkward to set up, so it is
rather striking that there are also two or possibly three cases in which *non-induced* donors
have elicited flowering in receptors of a different response group.

Both Lang and Zeevaart have found that
vegetative SDP Maryland Mammoth tobacco donors
in long days can induce flowering in defoliated
scions of LDP <u>Nicotiana</u> <u>sylvestris</u>, although
similarly defoliated but ungrafted plants of
<u>N</u>. <u>sylvestris</u> did not flower in long days. In
Zeevaart's experiments, eighteen out of thirty-
seven grafted scions flowered. Vegetative

Maryland Mammoth stocks in long days also induced some flowering in unvernalized biennial
Hyoscyamus scions. The third case, quoted by
Carr (1967) without supporting data, is that
leaves of LDP Sedum which were unable to support
flowering in defoliated scions of Sedum when returned to short days could nevertheless induce
flowering in SDP Kalanchoë.

Complementary Floral Stimuli

There is a ready explanation of these latter results, albeit an uncomfortable one for the florigen hypothesis. The explanation was first
considered by Lang in 1952 and rejected by him,
but it was subsequently adopted by Chailahjan.
Instead of one florigen there could be two complementary photoperiodic stimuli. One of these
would be more limiting in short day conditions,
particularly in long day plants, while the other
would be more limiting in long days, particularly in short day plants. Chailahjan identifies
the former of these with the gibberellins and
refers to the latter as anthesins. In some
plants, like the LSDP Bryophyllum, the content
of both stimuli needs to be boosted by the appropriate daylengths, in a given order, suggesting that they act on sequential stages of
the flowering process.
    According to Chailahjan's scheme, the
grafting of a scion in non-inductive daylengths
to a vegetative stock with the same daylength
requirements could never cause the scion to
flower because both stock and scion lack the
same component, gibberellins in LDP, anthesins
in SDP. However, a vegetative stock of one
response group could contribute the requisite
stimulus to a scion of the other response group
in non-inductive daylengths, as indicated
schematically on the next page.

	Florigen		Complementary stimuli		
Graft combination	Presence	Response	Gibberellins	Anthesins	Response
Receptor:					
LDP in short days	—	V	—	+	F
Donor:					
SDP in long days	—		+	—	
Receptor:					
SDP in long days	—	V	+	—	F
Donor:					
LDP in short days	—		—	+	

+, presence; -, absence
V, vegetative; F, floral

Schematic representations of the more usual graft combinations in terms of both the florigen and the complementary stimuli hypothesis would be:

	Florigen		Complementary stimuli		
Graft combination	Presence	Response	Gibberellins	Anthesins	Response
Receptor:					
LDP in short days	—	F	—	+	F
Donor:					
SDP in short days	+		+	+	
Receptor:					
SDP in long days	—	F	+	—	F
Donor:					
LDP in long days	+		+	+	

Note that Chailahjan's scheme does not account for the apparently more successful induction of flowering under these conditions, whereas the florigen hypothesis does so. Indeed, if gibberellins and anthesins are supposed to act in balance, such grafts should be less successful than those portrayed at the top of page 72. More graft combinations of plants with opposite photoperiodic requirements, each held in non-inductive conditions, need to be made before the relative merits of the florigen and complementary stimulus hypotheses can be assessed. The latter is more flexible, but almost too much so for a valid test to be made of it.

Photoperiodic Stimulus and Flower Hormone: Are They the Same?

Occam's razor should be used wherever possible in flowering physiology, where all generalizations soon encounter anomalies in one plant or another. Consequently, it was long assumed that the floral stimulus exported to the shoot apex from leaves after their exposure to inductive photoperiods was the same as the graft-transmitted flower hormone. Combined with the view that one flower hormone, florigen, was common to all plants, this assumption generated the problem of why florigen synthesis should have such different environmental requirements in different plants.

There is a further difficulty for this view, in that the leaves of some plants, e.g., Perilla, may become permanently inductive and able to initiate flowering in graft receptors long after they have been removed from the inductive daylength. Either we assume that synthesis of the floral hormone, initially controlled by daylength, eventually becomes

indifferent to it, or we must entertain the
possibility that the initial photoperiodic stimulus is different from the flower hormone exported by the permanently induced leaf.

   Although the leaves of Perilla can become
permanently inductive, as shown by their being
able to induce flowering in a whole series of
receptors to which they are successively
grafted, the individual plant does not remain
induced in non-inductive daylengths and soon re-
verts to vegetative growth. Presumably this is
because the permanently induced leaves are
superseded by younger non-induced leaves as the
major suppliers of assimilates to the shoot
apex.

   In Xanthium (SDP), Bryophyllum (LSDP) and
Silene (LDP) there is an even more striking expression of what has been called the induced
state. The presence of older, previously induced
leaves on a plant can result in the younger
leaves being induced indirectly, without any
exposure to inductive photoperiods, so that the
plant becomes permanently floral unless adverse
conditions destroy the induced state, as high
temperatures do in Bryophyllum (van de Pol,
1972). In fact, the whole plant becomes inductive and able to cause flowering in a graft receptor when neither partner is in an inductive
daylength. The receptor in turn is indirectly
induced and then is capable of acting as donor
of the induced state to a new vegetative
partner, ...and so on in a chain reaction.

   Consider Silene armeria in which flowering
can be induced by long days, by vernalization,
by short days at high temperatures, or by gibberellin treatment. Despite the variety of inducing conditions, and possibly therefore of
primary stimuli, Wellensiek (1966) was able to
show that all kinds of induced plants could act
as effective graft donors, even with their

leaves removed and in repetitive grafts. Similarly, Deronne and Blondon (1973) found that regardless of whether Perilla was induced to flower by exposure to short days or to low temperatures, its leaves could induce flowering in graft partners.

Several other lines of evidence suggest that such secondary or indirect induction may differ from primary photoperiodic induction in its processes and its products. Indirect induction may be confined to meristematic tissues, which is certainly not the case for photoperiodic induction. Perhaps related to this difference, van de Pol (1972) found that although juvenile leaves of Bryophyllum could not be photoperiodically induced, they could be indirectly induced by a graft partner. He also found that although Bryophyllum donors could induce flowering in Kalanchoë receptors, the latter could not undergo indirect induction like the donor.

The induced state is a fascinating phenomenon, with its implications of a semipermanent reprogramming of the genes. Its implications for the florigen hypothesis have not been fully thought through, but I believe there is much to be gained from further consideration of Carr's (1967) suggestion that the primary photoperiodic stimulus to flowering may not be the same as the contagiously propagated change transmitted by grafting.

Six

Clues to the Identity
of the Floral Stimulus

The reactions involved in the synthesis of the
floral stimulus by leaves remain a dark conti-
nent on biochemical maps, although not from
want of effort or ingenuity. Florigen has been
actively sought since it was conceived and named
forty years ago, and we are reaching the point
where the concept of a specific flower hormone
should itself be queried as much as our continuing failure to identify it.

Extraction and Analysis

The most decisive experiment would be to extract
florigen from photoperiodically induced plants
and by its application reproducibly and con-
vincingly cause flowering in plants under non-
inductive daylengths. Unfortunately, the results
when convincing have not been reproducible, and
when reproducible have not been very convincing.
There have been sporadic reports of quite strong
responses, beginning with the Bonner's extract
of young inflorescences of <u>Washingtonia</u> palms

and Roberts' extracts of induced <u>Xanthium</u>
plants, but duplication has proved difficult.

Procedures giving more repeatable results
have been devised by Lincoln and his colleagues.
Their extracts of induced <u>Xanthium</u> and sunflower
plants have consistently displayed some activity
when applied to vegetative <u>Xanthium</u> plants. Some
fractionation of this florigenic activity has
been achieved, leading Lincoln and Cunningham to
rename the active principle florigenic acid.
However, the flowering response obtained with
this extract is, at the most, minimal and is
soon lost as fractionation proceeds. Carr (1967)
subsequently reported that the response to such
extracts could be enhanced if gibberellins were
also applied. Hodson and Hamner confirmed this
for applications to <u>Xanthium</u>, which could be
construed as lending support to Chailahjan's
concept of complementary stimuli, but not for
applications to another SDP, <u>Lemna perpusilla</u>.
As so often in flowering physiology, idiosyncrasy asserted itself.

Analyses of leaves to see what changes in
metabolism take place during photoperiodic induction have frequently been made. The results
provide an embarassment of riches, changes
having been found in many of the components analyzed. Which changes are causal, which secondary, and which irrelevant, is difficult to
decide.

Applied Compounds and the Timing of Their Effects

Another approach towards identification of the
floral stimulus has been to apply compounds
active in other physiological processes and look
for effects on flowering. Salisbury developed
this technique into the most powerful tool we

have at the moment for exploring the nature of
photoperiodic induction. By using Xanthium
plants with only one leaf and requiring exposure
to only one short day, he had at his disposal a
reproducible system in which the effects of
varying sites (leaf or apex) and times of application of a compound could be related to the
sequential processes involved in flowering--
high intensity light process, timing reactions,
hormone synthesis, translocation, floral evocation, and development. A compound inhibiting
only hormone synthesis, for example, should have
little effect when applied at the end of the
dark period or thereafter, whereas a compound
inhibiting flower development should be most
inhibitory when applied soon after the dark
period and gradually become less so as flower
development proceeds. Figure 14 (next page)
illustrates the general form of Salisbury's results, while Figure 11 (page 59) illustrates a
specific example with the respiratory uncoupler,
2,4-dinitrophenol, giving results which imply
that this compound acts by inhibiting hormone
synthesis. Where an inhibitor acts on a specific
synthetic reaction, simultaneous application of
the product of that reaction should relieve the
inhibitory effect on flowering. Inhibitory effects of 5-fluorouracil on flowering should be
antidoted by simultaneous application of uracil
or orotic acid, but not by thymidine, if it is
the suppression of RNA synthesis that reduces
the flowering response. Natural plant growth
substances, and inhibitors of their synthesis,
may also have their role in the flowering process analyzed in this way. Salisbury has done
many experiments of this kind with Xanthium
which are reviewed in his book, while others
have used Pharbitis and Chenopodium. Among LDP,
only Lolium has been used to any extent in experiments of this kind.

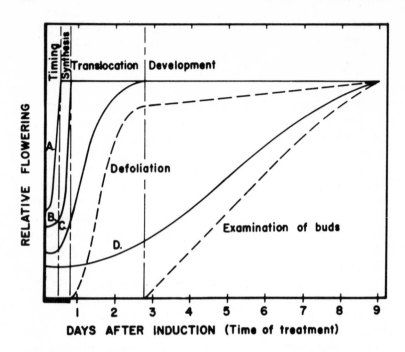

Figure 14. Schematic representation of the effect on flowering response of the time of application of compounds which influence time measurement (curve A), hormone synthesis (curve B), translocation of the hormone (curve C) or development of the inflorescence (curve D) (Salisbury, 1963).

On the other hand, a great many plants have been subjected to "look-see" experiments in which potentially active substances have been applied to them in the hope of eliciting a flowering response. The current enthusiasms of plant physiology shape the choice of substances. Auxins were the first to be used extensively, followed by gibberellins and inhibitors of nucleic acid and protein synthesis, cytokinins, abscisic acid, ethylene, and (recently) cyclic

AMP and acetylcholine, all of which can affect flowering substantially in some plants.

(a) <u>Auxins</u>. Auxins and their antagonists figured largely in early experiments, particularly after 1942 when auxins were found to promote flowering in pineapples and tri-iodobenzoic acid was found to promote it in soybeans and tomatoes. In general, auxins tended to inhibit flowering in short day plants but to promote it to some extent in LDP such as <u>Hyoscyamus</u>, <u>Silene</u>, and <u>Lolium</u> under marginal conditions. Their action therefore had a symmetry suggestive of participation in photoperiodic reactions but not, of course, of florigen which should be promotive for both long and short day plants. This latter problem could be avoided by the argument that endogenous auxin concentrations were suboptimal in long day plants and needed to be raised, whereas in short day plants they were higher and additional auxin made them super-optimal and therefore inhibitory. In fact, this became quite a tidy theory, able to explain many observations such as the promotive effect of anti-auxins on flowering in SDP. But fashion moved on, and Salisbury showed that the inhibitory effect of auxin was exerted not only during the inductive dark period but also during the subsequent period of translocation. Thus, auxin effects on flowering could not be due solely to action on the photoperiodic processes in the leaves.

(b) <u>Gibberellins</u>. The next team up, the gibberellins, has come closest to matching our aspirations for them. Their application to plants in short days has caused flowering in more than twenty different LDP and has replaced the requirement for vernalization in another twenty. The endogenous level of gibberellins is higher in long days, or after light breaks, in

both LDP and SDP, and the application of substances inhibitory to gibberellin biosynthesis
can inhibit flowering in several long day
plants. Baldev and Lang's (1965) experiments
with the LDP Samolus parviflorus provide an
elegant example. Application of gibberellic
acid (GA3) to plants in short days induced
flowering. Application of Amo-1618 or CCC, which
inhibit gibberellin synthesis, to plants in long
days prevented them from flowering. The more
Amo-1618 applied and the fewer the number of
long days given, the greater was the amount of
GA3 required to restore flowering.

Unfortunately, this consistent pattern of
results has not been found with several other
LDP. In Lolium, for example, application of GA3
induces flowering in short days, but application
of CCC has no effect on the flowering of plants
exposed to one long day and even increases the
response to applied GA3 in both long and short
days. Even more different are the many LDP,
particularly those with a caulescent (i.e.,
non-rosette) habit of growth, which do not
flower after even massive and repeated applications of gibberellins. In some cases the most
effective gibberellin may not have been applied,
for their relative effectiveness in causing
plants to flower varies greatly between species.
But there are even long day plants, such as
Fuchsia and Proserpinaca, whose flowering is inhibited by the application of gibberellins.

With the LSDP Bryophyllum Zeevaart and
Lang, in a series of papers, have clearly implicated gibberellins as an important component of
the initial long day reactions. Applied gibberellins can replace the need for long days to
the extent that not only do the plants flower
in short days, but their leaves can induce
flowering when grafted to other plants. Flowering in Bryophyllum is suppressed by

application of CCC, which inhibits gibberellin synthesis. Gibberellin A_{20} appears to be the effective endogenous gibberellin in this case, and Zeevaart has shown that it acts in the leaves, even in the absence of buds and is not detectable in leaves in short days unless these have been preceded by several long days (Zeevaart, 1973).

In the SLDP Coreopsis grandiflora, GA3 can replace the requirement for long days.

A great variety of results has been obtained with SDP. Gibberellins promote flowering in some, such as Chrysanthemum morifolium and Perilla under suboptimal daylength conditions. Applied before the critical night length has been reached, gibberellins may increase flowering in Xanthium and Pharbitis. They also enhance the flowering response to extracts of induced plants applied to Xanthium. On the other hand, gibberellins may inhibit flowering in other SDP such as Kalanchoë, Chenopodium, and strawberry. Inhibitors of gibberellin synthesis may promote flowering in some of these plants but not in others.

Clearly, gibberellins play a key role in the induction of flowering in many plants and can be promotive for plants in all photoperiod response groups, as envisaged in Chailahjan's hypothesis of complementary photoperiodic stimuli. This supposes that gibberellin levels are adequate in SDP even in short days but are only raised to the requisite levels in long day plants by exposure to long days. However, Grigorieva et al. (1971) found, on comparing gibberellin levels in Nicotiana sylvestris (LDP) and Maryland Mammoth tobacco (SDP), that although long days raised the level in both species, it was much higher in either daylength in the LDP. In fact, gibberellins could not be detected in the SDP in short days, a result

difficult to reconcile with Chailahjan's
hypothesis.

(c) <u>Cytokinins</u>. Cytokinin application has pro-
moted flowering in several SDP such as <u>Perilla</u>,
<u>Chenopodium</u>, <u>Pharbitis</u>, and <u>Wolffia</u>, and even in
one LDP, <u>Arabidopsis</u>. In <u>Pharbitis</u> the promotion
by kinetin was most pronounced when it was applied to leaves during the first four hours of
the inductive dark period, which suggests that
it was influencing the photoperiodic processes
in the leaves.
    Endogenous levels of cytokinin were shown
several years ago to be higher in <u>Begonia</u> leaves
in short days, and this effect of daylength has
recently been confirmed for buds and young
leaves of <u>Xanthium</u> (Staden and Wareing, 1972)
and for the xylem sap of <u>Perilla</u> (Beever and
Woolhouse, 1973). At least in <u>Perilla</u>, however,
the increase in cytokinin level seems to take
place after rather than during photoperiodic
induction.

(d) <u>Abscisic acid</u>. Abscisic acid, like the other
endogenous growth substances, appears to be influenced by daylength. Several early experiments
suggested that it is more abundant in short
days, but the generality of this relation is
currently in some doubt. When supplied exoge-
nously it has been found to inhibit flowering
in several LDP and to induce flowering in a
number of SDP in long days, as in <u>Pharbitis</u>,
<u>Chenopodium</u>, <u>Ribes</u>, and strawberry. But the
timing of its inhibitory effect in <u>Lolium</u> suggests that it acts at the shoot apex during
floral evocation rather than in the leaf.

(e) <u>Ethylene</u>. Fires nearby have long been known
to cause flowering in pineapple fields, an
effect now ascribed to a local increase in the

ethylene content of the air. Ethylene release is now used commercially to promote flowering in pineapples, and may promote it in other plants too, such as Plumbago indica (SDP). With Pharbitis, on the other hand, ethylene is a potent inhibitor of flowering, which Suge (1972) has shown to act in the cotyledons and only during the inductive dark period.

(f) Steroids and polypeptides. Since many animal hormones are either steroids or polypeptides, there has been a variety of attempts to see whether the flower hormone is a related compound. Estrogens were a natural early hope, and daylength has been shown to influence their occurrence in plants. Kopcewicz (1972a,b) found estrogen levels to rise sharply in SDP Perilla and Chenopodium in short days, and in LDP Hyoscyamus and Salvia in long days. In all cases the rise was most rapid at about the time of flower initiation. Thus, the changes have the requisite symmetry and timing for the flower hormone. Inhibitor studies have also implicated steroid synthesis in the process of photoperiodic induction in both long and short day plants. Bonner et al. (1963) found that an inhibitor of zymosterol synthesis, SK & F 7997, inhibited flowering in two SDP, Xanthium and Pharbitis, when applied to leaves near the beginning of the dark period. It also strongly inhibits flowering in Lolium (LDP), acting in the leaf with a timing (see Figure 15, next page) which suggests that steroid synthesis is a component of photoperiodic induction. Numerous attempts have been made to induce flowering by the application of a variety of steroids with occasional but not striking success, and there the matter rests, tantalizingly.

Evidence for a polypeptide flower hormone is less convincing. Inhibitors of protein

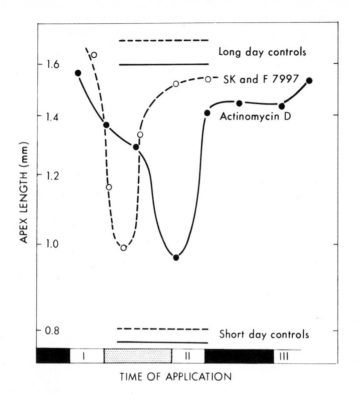

Figure 15. Effect on flowering response in
Lolium of applications at various times during
exposure to one long day of the steroid inhibitor SK & F 7997 to leaves and of Actinomycin D
near the shoot apex (Evans, 1969).

synthesis can suppress the flowering response,
but in most cases this seems to be due to action
at the shoot apex rather than in the leaf. In
Xanthium, however, p-fluorophenylalanine and
cycloheximide can prevent leaf induction, but
their action may not have been on protein

synthesis: phenylalanine, for example, is a precursor of many aromatic compounds.

There is no evidence to suggest that new RNA synthesis in the leaves of LDP is needed during photoperiodic induction, but this possibility cannot be ruled out for at least some SDP such as Pharbitis, Chenopodium, and rice. Application of inhibitors of RNA synthesis during photoperiodic induction suppressed flowering in these plants, but action at the shoot apex rather than in the leaf is not excluded. However, a change in the base ratios of new messenger RNA in Pharbitis cotyledons exposed to a long dark period has been detected.

Taken as a whole, these results indicate that there are several groups of compounds whose concentration in leaves is influenced by daylength, and which can intervene to tip the balance from vegetative growth to flowering under some conditions. The gibberellins are preeminent among these, but steroids, abscisic acid, auxins, cytokinins, and ethylene may all participate. These effects have been dismissed by some flowering physiologists as unspecific side effects, but where then is the master florigenic substance? Perhaps only in the mind's eye.

Seven

What Happens at the Shoot Apex?

For many years the shoot apex was the Cinderella of flowering physiology, most attention being focused on the leaf reactions. The one term, photoperiodic induction, was used to cover processes in both leaf and apex. Because they are likely to be very different in nature, the term induction should be reserved for processes in the leaf, while those at the shoot apex following the arrival of the floral stimulus and leading to the initiation of flower primordia can be referred to as evocation. It is these processes that concern us now.

We have seen that many changes occur in leaves under inductive photoperiods, several of which are likely to be communicated to the apex in due course. Defoliation experiments tell us when the last or slowest-moving of the substances causing floral evocation has been exported from the leaf after a single inductive cycle. This, together with some measure of the speed of movement, allows us to estimate fairly closely when the photoperiodic stimuli reach the shoot apex. In both <u>Lolium</u> and <u>Pharbitis</u> this is likely to be during the morning after the inductive day. That is the time, therefore, at which to look for the earliest processes of floral evocation. But before doing so, let us consider the characteristics of the vegetative shoot apex, the field on which the photoperiodically influenced stimuli and inhibitors must interact.

The Vegetative Apex

Contrary to widely held notions, the vegetative shoot apex is a relatively inactive organ, with only about two percent or less of its nuclei in mitosis at any one time, a far lower proportion than in root apices. The ratio of RNA to DNA in the vegetative shoot apex is also low and may even be less than one. To some extent this may be due to an accumulation of nuclei in what is known as the G_2 stage, which means they have doubled their DNA content in preparation for mitosis but have not yet divided. Bernier and his colleagues at Liège have shown that about half of the nuclei in the vegetative shoot apex of Sinapis are in the G_2 stage, but such a high proportion may not occur in other photoperiodic plants.

The other factor tending to lower the RNA/DNA ratio in vegetative shoot apices is the occurrence of a group of cells at the summit of the apex which histochemical staining shows to be extremely low in RNA, as may be seen in Figure 16a (page 91). Long known in dicotyledonous plants, this inactive "central zone" was considered by the French anatomist Buvat to be a waiting meristem, méristème d'attente, awaiting the message from induced leaves to go ahead and form flowers. However, this inactive central zone is not apparent in the apices of many monocotyledonous plants, such as Lolium. Also, the longer that dicotyledonous plants are kept in non-inductive daylengths, the more RNA accumulates in the

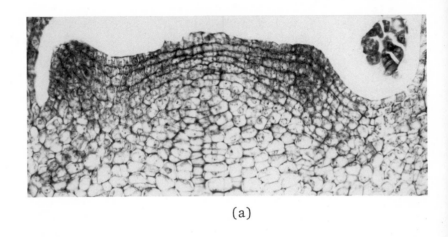

(a)

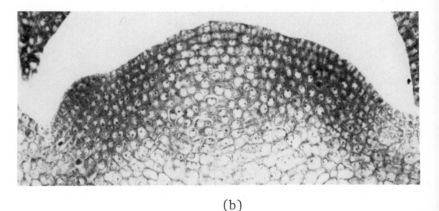

(b)

Figure 16. Vegetative (a) and florally evoked (b)
apical meristems of Sinapis alba, stained with
Azure B. Note the low stain intensity in the
inactive central zone, and the more dome-shaped
and layered structure of the evoked meristem,
fixed 42 hours after the beginning of the long
day exposure (Bernier, 1971).

central zone, until its staining intensity approaches that of the surrounding peripheral zone which gives rise to leaf primordia. Such apices are described by Nougarède as being in the intermediate stage, but they remain vegetative in the absence of the inductive stimulus.

Evocation: RNA Synthesis

Following the arrival at the apex of the inductive stimulus, many new genes must eventually be called into play to determine floral characteristics, but we have no evidence that derepression of floral genes and the synthesis of novel messenger RNA and proteins occur early in evocation. No one has yet succeeded in detecting them, but given the small size of the shoot apex, a few micrograms in weight at the most, this may be due to a lack of sufficiently sensitive techniques.

Novel RNA may or may not be required, but RNA synthesis certainly is. Inhibitors of RNA synthesis, such as Actinomycin D and 5-fluorouracil, can suppress floral evocation when applied to the shoot apex at about the time of arrival of the photoperiodic stimulus (see Figure 15). Comparable effects with 5-fluorouracil can be reversed by application of orotic acid or uracil at the same time or within a few hours. With the rather large shoot apices of Lolium it has also been possible to show a rise in ^{32}P incorporation into RNA during the morning after the long day. This appeared to be in all fractions of RNA, but microautoradiographic studies of evocation in Sinapis suggest that the early increase in RNA synthesis is more pronounced in the chromatin than over the nucleolus.

An interesting problem is posed by several experiments with inhibitors of RNA synthesis applied to the shoot apex. With Xanthium, for

example, they are inhibitory only when applied to
the apex before the middle of the inductive dark
period, before the floral stimulus has begun to
be exported. With Sinapis, also, applications to
the apex of the inhibitor 2-thiouracil are most
inhibitory after only twelve hours of light,
i.e., at about the time the critical daylength is
reached but before the floral stimulus could have
reached the apex. Later applications are not inhibitory. Similar results have been obtained with
Anagallis. They suggest either that there is a
substantial delay in inhibitor uptake or that
some preparatory synthesis of RNA at the shoot
apex is needed before the arrival of the floral
stimulus. Indeed, Gressel, Zilberstein, and Arzee
found a transient rise in uridine incorporation
into RNA in Pharbitis buds at about the tenth
hour of the inductive dark period which could be
prevented by a light break.

These results suggest either that there is a
preliminary message from the leaves to the apex,
in advance of the subsequent arrival of the inductive stimulus, or that the bud itself is also
responding directly to the inductive photoperiod
(see p. 27). This latter alternative could explain why Lolium differs from the other plants in
not displaying such early inhibitory effects, because its equivalent of the apical bud is con-
cealed inside many leaf sheaths and precluded
from receiving light-dark signals directly.

Given that a rise in RNA synthesis at the
apex is one of the earliest essential steps in
floral evocation, what is its location? Does it
occur only in Buvat's waiting meristem, or is it
more general? Only in four plants have the histochemical and autoradiographic studies been de-
tailed enough to answer the question, namely
Sinapis, Lolium, Pharbitis, and Perilla. In all
four the answer is the same: the increase is
greatest in the previously inactive central zone,

but it is also pronounced in the previously active peripheral zone. In Perilla the increase in
total RNA is paralleled by increase in numbers of
ribosomes.

Thus, evocation does not consist simply of
activation of a méristème d'attente as Buvat
imagined. Rather, there is an initial elimination
of the vegetative zonation coupled with an activation of RNA synthesis over the whole surface of
the apex. In fact, even young leaves and leaf
primordia near the shoot apex tend to increase
their incorporation of RNA precursors.

Histones and Other Proteins

Experiments with Lolium reveal an increased incorporation of precursors into protein as well as
into RNA in the apex at the time of arrival of
the floral stimulus. Subsequent changes in protein content and distribution run parallel to
those in RNA in all the plants examined, except
in the case of the histones.

Rather dramatic changes in histone contents
of Xanthium and Chenopodium apices during evocation were reported by Gifford in 1963. There was
a sharp fall in the intensity of nuclear staining
and a rise in cytoplasmic staining, originally
interpreted as indicating extensive derepression
of "floral genes" by loss of their histones in
the process of evocation. Our current interpretation of these results is rather different. An
early effect of evocation, but by no means the
earliest in Lolium, is a swelling of the nuclei
of the previously inactive cells of the apex.
This dilutes the histochemical staining of DNA
and does the same for associated histones. At the
same time there is a pronounced increase in the
number of ribosomes, whose basic proteins react
with the Fast Green stain for histones, giving an

apparent increase in cytoplasmic histone. Subsequent histochemical studies have not demonstrated any marked change in histone content relative to DNA in apices at evocation (Jacqmard et al. 1972), but in Lolium we did find evidence of a possible shift in the composition of the histones.

DNA Synthesis and Mitosis

Activation of RNA synthesis by the apex at evocation is associated with a pronounced swelling of the nuclei, probably analogous to loop formation and the swelling of polytene chromosomes when genes are derepressed for RNA synthesis. Thus, it is of considerable interest that Nagl (1973) has found the activity and structure of the polytene chromosomes of the suspensor cells in Phaseolus embryos to be under photoperiodic control. DNA synthesis does not appear to be needed for floral evocation in Xanthium, Lolium, or Sinapis; in fact, its inhibition at the apex by applications of 5-fluorodeoxyuridine immediately after an inductive dark period may even promote flowering in Chenopodium. In this context it is Pharbitis which asserts its idiosyncrasy, in that inhibitor application experiments by Zeevaart make it clear that DNA synthesis in the shoot apex is essential at about the time of arrival there of the floral stimulus.

In most active root and shoot meristems so far examined the majority of nuclei are in the G_1 phase (Jacqmard et al. 1972) and must therefore undergo DNA replication before they enter mitosis. Sinapis is perhaps exceptional in this respect in that at least half the nuclei in a responsive vegetative apex are in the post-synthetic G_2 phase, and many of these enter mitosis at about the time of arrival of the floral stimulus. Bernier (1971) attaches considerable significance

to this effect, since evocation of flowering in
Sinapis by whatever means is always accompanied
by an early increase in mitosis at the shoot
apex, which he sees as essential to the reprogramming of the genome for floral morphogenesis.
However, applications of colchicine, which blocks
mitosis at metaphase, do not inhibit floral evocation in Sinapis and, moreover, there is other
evidence that the stimulation of mitosis is separate from the flowering process in that it can
occur under daylengths too short to induce flowering. Also, no plant other than Sinapis offers
any evidence of mitosis being a prerequisite for
floral evocation. For the floral morphogenesis
that ensues, mitosis and DNA synthesis are obviously required in order to provide the additional cells, but probably not for evocation in
most plants.

The Geometry of Floral Evocation

Our present knowledge of early events at the
shoot apex following the arrival of the floral
stimulus does not permit us to differentiate between two very different views of what happens.
On the one hand, it is envisaged that a specific
flower hormone, florigen, derepresses the floral
genes in the cells of the apex, leading to the
synthesis of novel messenger RNA and eventually
protein. This still leaves us with the problem
of how the new three-dimensional structures of
the flower are specified, but that is a problem
shared with all biology. Such an interpretation
of floral evocation leans heavily on the dogmas
of molecular biology and is probably favored by
most flowering physiologists in spite of a complete lack of evidence for the appearance of
novel RNA or protein in the shoot apex at
evocation.

Alternatively, evocation may be viewed as resulting from a general activation of the shoot apex, leading to an increase in its size and elimination of its vegetative pattern of activity, thereby allowing a new floral geometry to be established. At this somewhat later stage, obviously, new genes must be brought into play in order to specify the many kinds of inflorescence and flower structure.

Before dismissing this apparently naive explanation, pause to consider some of its virtues. For example, it makes it far easier to explain how plants like Silene can have several different environmental pathways to floral evocation; how so many different growth substances and other compounds can elicit flowering in non-inductive conditions; how photoperiodic inhibitors can interact with photoperiodic stimuli; how a photoperiodic stimulus can influence other processes, such as tuberization or germination, as well as flowering; how the stimulus to floral evocation can also influence the rate and extent of differentiation of the various organs of the flower; or how roots, stems, and young leaves can have a determinative effect on flowering in some plants.

One of the most striking early effects of floral evocation in many plants is that the shoot apex swells and becomes turgid, which could be due to an increase in the permeability of the apical cells or in their competitive ability to import assimilates and solutes. As a result of such swelling, dicot apices become larger and dome-like, and their cellular structure becomes more layered and regular (see Figure 16). Increased RNA and protein synthesis, especially in the central zone, results in the development of a more uniformly active surface. If we assume, as suggested by Snow and others, that primordia require a certain apical space and distance from earlier primordia before they can be initiated,

it follows that an enlarged apical surface with a
greater level of activity should permit a changed
pattern of organ formation, as required for flower initiation. The new pattern, of course, will
be somewhat influenced by the vegetative pattern
since the leaves supplying the shoot apex will
generate gradients in the concentration of assimilates and growth substances which reflect leaf
arrangement and vascular connections.

Increased apical dimensions and activity
could also influence the degree of apical dominance over lateral bud development. The essence of
flower initiation is more rapid and simultaneous
development of leaf-like primordia to give spi-
rals or whorls of sepals and petals, combined
with precocious initiation of axillary buds. In
the grasses and cereals, for example, these latter give rise to panicle branches, spikelets,
florets and stamens, and in dicots to inflorescence branches and stamens.

Heteroblastic plants, in which leaf shape
changes with progress towards flowering, may provide a good example of the dependence of evocation on shoot apex geometry. For at least several
of these plants, such as Cannabis and Sinapis,
there is evidence that the change in leaf shape
reflects changing apical organization. The rate
of this change depends on temperature and daylength. In Vicia faba, for example, the number of
leaflets in each leaf increases progressively at
successive nodes, at a rate which is faster the
longer the day and the lower the temperature.
Progress towards flowering is similarly affected
by these environmental factors, so that flower
evocation always coincides with the attainment of
a particular number of leaflets and presumably,
therefore, with a particular structure of the
shoot apex.

I have concentrated discussion on this
apical geometry view of evocation rather than on

the more fashionable floral operon derepression model because it tends to be ignored and because it can be helpful in explaining quite a number of the findings which are so awkward for the floral operon model. In doing so, however, I have not meant to imply that the latter may not also be operative, and there is no reason why both mechanisms may not contribute to the evocation of flowering. They are not mutually exclusive.

Eight

Flower Development

For many of the plants used in flowering experiments, photoperiodic treatments which lead to floral evocation usually suffice for full flower development, even in plants requiring only one inductive cycle. However, exposure to additional inductive cycles often causes flower development to proceed more rapidly (see Figure 5). This is true even when the additional inductive cycles are given several days after evocation has occurred and floral morphogenesis has begun.

In other plants, such as Anagallis, only a few inductive cycles are needed for evocation, but more are required for flower development to be completed, as Garner and Allard found in their early experiments with soybeans. Cynosurus cristatus, crested dogstail grass, requires exposure to only three long days for floral evocation, but normal inflorescence development requires at least fourteen long days. With less than that, floret morphogenesis is incomplete, and leafy shoots may form in the inflorescence. Such a response, which is found in other grasses too, can be of adaptive value in cold wet climates because the inflorescence plantlets so

formed in autumn can establish themselves independently before winter sets in. Lack of
sufficient inductive cycles for full flower development may cause floral morphogenesis simply
to stop at some point, but in other cases there
may be actual reversion to the vegetative pattern
at the shoot apex. Yet other plants, like
Caryopteris clandonensis, are indifferent to daylength for floral evocation but have strict requirements for floral morphogenesis.

What are we to assume when one long day suf-
fices for evocation but additional long days,
even well after evocation has occurred, acceler-
ate floral morphogenesis or are needed to complete it? If florigen acts by derepressing an
early gene in the floral operon, why does an additional long day help when floral morphogenesis
is well under way? Consider the formation of a
grass inflorescence. In essence the first step
is precocious initiation of axillary buds so that
instead of forming tiller branches much later
they form panicle branch or spikelet primordia
while still close to the shoot apex through a
temporary reduction of apical dominance. The
spikelet primordia must then in turn undergo
precocious axillary bud development when, having
initiated one or two leafy sterile glumes, they
then form floret primordia, and the cycle is repeated again at the initiation of anthers. Thus,
a particular kind of influence, a suspension of
apical dominance, must be applied to the shoot
apex not only for the evocation of flower primordia but also for their subsequent development.
It is easy to see, therefore, why crested dogstail grass needs those additional long days to
prevent reversion to leafy inflorescences, or
how Lolium inflorescences can respond to additional long days by developing faster.

In the short day plant Kalanchoë, likewise,
minimal evocation involves suppression of the

terminal growing point and precocious release of
the uppermost two laterals. Additional short days
repeat this cycle, each resulting in an additional order of branching and a doubling of the number of flowers (see Figure 3). Such responses are
quite compatible with the view that floral evocation involves changed apical correlations.

Alternatively, one may envisage that floral
morphogenesis proceeds through a series of phases
each determined by a set of genes and each bring-
ing the next set in the floral operon into action.
Once initiated by floral evocation the process is
canalized to proceed to completion, as appears to
be the case for some plants. Heslop-Harrison
(1963) has presented a model of this kind for
floral morphogenesis.

Apical Surgery and Its Interpretation

Experiments in which floral primordia are bisected at various stages of their development are
of interest in this context. The earliest were
those of Cusick (1956) with Primula bulleyana,
subsequently extended to Portulaca (Soetiarto and
Ball, 1969), Nicotiana (Hicks and Sussex, 1971),
and Aquilegia (Jensen, 1971). Considerable skill
is required to operate successfully on the very
small, elastic and inaccessible primordia, but
the results with these very different plants
agree in their main feature. For all four plants
the later the stage at which the floral primordia
were bisected, the fewer were the kinds of organs
that could be regenerated near the cut surface.
Replacement sepals were regenerated only when
flower primordia were bisected well before the
appearance of sepal primordia. More advanced
floral primordia could initiate petals, stamens,
and carpels, but not sepals. Still more advanced
ones could initiate carpels, but not petals and

stamens. The primordia which formed above the cut
surface were always of the kind and at the same
stage as those forming away from the cut.

Thus, as Cusick (1956) concluded, the floral
meristem appears to pass through a succession of
physiological states that permit and regulate the
formation of each kind of organ in turn. These
could be due to irreversible progression through
a series of genetically programmed states each
switched on by an inducer from the preceding
round of morphogenesis, as Heslop-Harrison (1963)
suggested. He envisaged that these inducers from
the earlier-formed floral organs would operate
over a short intercellular range, inducing formation of the next kind of floral organ nearby.

However, the primordia bisection experiments with all four plants agree in showing that
later-formed floral organs, such as petals and
stamens, may be initiated above the cut surfaces
in spite of the absence of sepals there. Thus,
the formation of a complete whorl of stamens or
petals is not dependent on positional information
or localized induction from a preceding and complete whorl of sepals, a conclusion supported by
studies of abnormal flowers of many kinds. If an
inducer of petal-formation is generated during
sepal initiation, as Heslop-Harrison supposes, it
clearly does not act locally at the nearest petal
site, but rather at the floral meristem.

Jensen's (1971) continuous photographic rec-
ord of what happens in bisected Aquilegia buds is
highly suggestive in this context, because he
found that floral organs were initiated above a
cut surface only after a new apex was regenerated
in the bisected half, although initiation on the
side away from the cut could proceed before that.
Clearly, expression of the genetic pattern of
floral organs depends more on information from
the apex than on that from leaf and early-formed
floral primordia.

Sex Expression

Even at quite late stages in floral morphogenesis,
its pattern is still modifiable by environmental
conditions and growth substance application, as
is notably evident in the sex expression of flowers. In Cannabis sex determination occurs soon
after flower initiation and the flower primordium
is basically unisexual throughout its development, but short days and auxins increase the proportion of primordia which form female flowers.
In other plants such as maize and several cucur-
bits, the floral primordium remains potentially
bisexual for some time after its inception, the
switch to one sex or another being determined by
the genome in interaction with daylength, temperature, auxin levels, and gibberellin levels. In
general, short days, low temperatures, and the
application of auxins, ethylene, or antigibberel-
lins promote femaleness whereas gibberellins
promote maleness.

It is not only the sex organs of flowers
whose development may be controlled by daylength,
as Garner and Allard found in their early experiments which included a species of Viola. Violets
in short days, like those in spring, produce
flowers with purple petals whereas those in long
days, like plants in summer, produce only cleis-
togamous flowers without petals. The breeding
system of plants may be influenced by daylength
in other ways too. Florets of the more polyploid
races of kangaroo grass, Themeda australis, may
be mainly sexual or mainly apomictic (i.e., non-
sexual) depending on the daylength. Regardless
of the daylength requirements for flower initiation, short days increased asexual apospory,
while long days favored sexual reproduction.

Thus, daylength exercises substantial control of both the rate and pathway of floral
morphogenesis as well as of the earlier processes

of evocation, interacting with the genome and
modifying it to some extent via control of the
breeding system.

Conclusion

Molecular biology, with its hypotheses of such universal relevance and acceptance that they are nicknamed dogmas, is often taken as the paradigm for all biology. By comparison, the physiology of flowering is not at first base in terms of the generality and explanatory power of its major hypotheses, as any reader who has survived the preceding pages will agree.

But there is another face to biology, complementary to that concerned with the replication, transcription, and translation of genetic information. It is concerned with the myriad and subtle ways in which organisms adapt to their environments. Here we must expect variety rather than universality of response, particularly in plants because they can neither migrate nor hide from adverse weather. In the course of a long evolutionary history in a diversity of niches, plants are bound to have exploited an enormous variety of controls on their reproductive behavior. Faced with this the molecular biologist might throw up his hands in horror, but part of the fascination of flowering physiology lies in coming to understand not only the mechanisms but also the adaptive value of particular responses.

There is as great a danger in making a vir-
tue of variety, however, as there is in conceal-
ing or ignoring it. General principles must be
sought, but those so far devised, whether for
time measurement by phytochrome reversion or for
a universal flower hormone, seem accident-prone
because of the variety of alternate pathways to
flowering.

Like any field of research which is explored
in depth and with variety of attack, the physiology of flowering leads to and offers insights into many of the major problems in biology. Perhaps
the greatest of these is how a three-dimensional
structure as complex as a flower can be specified
genetically and formed so reproducibly. Another
is how time can be measured so accurately, the
essence of which still seems to evade us. A great
deal has already been accomplished, as I hope the
preceding pages have indicated, but it is the
problems still to be explained, like the moun-
tains still to be climbed, that seem most fascinating. There is little danger of them becoming a
non-renewable resource.

General References

Bernier, G. ed. 1970. Cellular and molecular aspects of floral induction. pp 492. Longman, London.

Chouard, P. 1960. "Vernalization and its relations to dormancy." Ann. Rev. Plant Physiol. 11:191-238.

Evans, L. T. ed. 1969. The Induction of Flowering. pp 488. Macmillan, Melbourne, London.

Evans, L. T. 1971. "Flower induction and the florigen concept." Ann. Rev. Plant Physiol. 22: 365-394.

Hillman, W. S. 1963. The Physiology of Flowering. pp 164. Holt, Rinehart, Winston, New York.

Lang, A. 1965. "Physiology of flower initiation." Handb. Pflanzenphysiol. XV/1, 1380-1536.

Salisbury, F. B. 1963. The Flowering Process. pp 234. Pergamon, Oxford.

References Cited

Many of the earlier experiments referred to in the text have not been given specific citations, in an attempt to keep this bibliography to reasonable proportions. However, references to all these uncited experiments may be found in my two earlier reviews mentioned in the general references on the previous page.

Baldev, B., Lang, A. 1965. "Control of flower formation by growth retardants and gibberellin in Samolus parviflorus, a long day plant." Amer. J. Bot. 52: 408-417.

Beever, J. E., Woolhouse, H. W. 1973. "Increased cytokinin from root system of Perilla frutescens and flower and fruit development." Nature New Biol. 246: 31-32.

Bernier, G. 1971. "Structural and metabolic changes in the shoot apex in transition to flowering." Can. J. Bot. 49: 803-819.

Bonner, J., Heftmann, E., Zeevaart, J. A. D. 1963. "Suppression of floral induction by inhibitors of steroid biosynthesis." Pl. Physiol. 38: 81-88.

Borthwick, H. A., Hendricks, S. B., Parker, M. W. 1948. "Action spectrum for photoperiodic control of floral initiation of a long day plant, winter barley (Hordeum vulgare)." Bot. Gaz. 110: 103-118.

Borthwick, H. A., Hendricks, S. B., Parker, M. W. 1952. "The reaction controlling floral initiation." Proc. Natl. Acad. Sci. 38: 929-934.

Butler, W. L., Hendricks, S. B., Siegelman, H. W.
    1964. "Action spectra of phytochrome in
    vitro." Photochem. Photobiol. 3: 521-528.

Carr, D. J. 1967. "The relationship between
    florigen and the flower hormones."
    New York Acad. Sci. Ann. 144: 305-312.

Chorney, W., Rakosnik, E., Dipert, M. H.,
    Dedolph, R. R. 1970. "Rhythmic-flowering
    response in Cocklebur." Bioscience 20:
    31-32.

Cumming, B. G., Hendricks, S. B., Borthwick,
    H. A. 1965. "Rhythmic flowering responses
    and phytochrome changes in a selection of
    Chenopodium rubrum." Can. J. Bot. 43:
    825-853.

Cusick, F. 1956. "Studies of floral morphogene-
    sis. I. Median bisections of flower
    primordia in Primula bulleyana Forrest."
    Trans. Roy. Soc. Edinb. 63: 153-166.

Cusick, F. 1959. "Floral morphogenesis in
    Primula bulleyana Forrest." J. Linn. Soc.
    Bot. 56: 262-268.

Deronne, M., Blondon, F. 1973. "Transmission,
    par greffe de feuilles, de l'induction
    florale acquise par l'action d'une basse
    température (5°C) en dyspériodes chez le
    Perilla ocymoides L., plant de jours courts.
    Comparaison avec la transmission de
    l'induction provoquée par les jours courts."
    CR Acad. Sci. (Paris) 277 Ser D: 1013-1016.

Downs, R. J. 1956. "Photoreversibility of flower
    initiation." Pl. Physiol. 31: 279-284.

Evans, L. T. 1960. "Inflorescence initiation in
    <u>Lolium</u> <u>temulentum</u> L. II. Evidence for inhib-
    itory and promotive photoperiodic process-
    es involving transmissible products."
    <u>Aust. J. Biol. Sci</u>. 13: 429-440.

Evans, L. T., King, R. W. 1969. "Role of phyto-
    chrome in photoperiodic induction of
    <u>Pharbitis</u> <u>nil</u>." <u>Zeitschr. f. Pflanzen-
    physiol</u>. 60: 277-288.

Evans, L. T., Wardlaw, I. F. 1966. "Independent
    translocation of ^{14}C-labelled assimilates
    and of the floral stimulus in <u>Lolium
    temulentum</u>." <u>Planta</u> 68: 310-326.

Fredericq, H. 1964. "Conditions determining
    effects of far red and red irradiation on
    flowering response of <u>Pharbitis</u> <u>nil</u>."
    <u>Pl. Physiol</u>. 39: 812-816.

Garner, W. W., Allard, H. A. 1923. "Further
    studies in photoperiodism, the response of
    the plant to relative length of day and
    night." <u>J. Agr. Res</u>. 23: 871-920.

Gibby, D. D., Salisbury, F. B. 1971. "Participa-
    tion of long day inhibition in flowering of
    <u>Xanthium</u> <u>strumarium</u> L." <u>Pl. Physiol</u>. 47:
    784-789.

Gressel, J., Zilberstein, A., Arzee, T. 1970.
    "Bursts of incorporation into RNA and
    ribonuclease activities associated with
    induction of morphogenesis in <u>Pharbitis</u>."
    <u>Devel. Biol</u>. 22: 31-42.

Grigorieva, N. Y., Kucherov, V. F., Lozhnikova,
    V. N., Chailahjan, M. K. 1971. "Endogenous
    gibberellins and gibberellin-like

substances in long-day and short-day species of tobacco plants: a possible correlation with photoperiodic response." Phytochem. 10: 509-517.

Hamner, K. C. 1940. "Interrelation of light and darkness in photoperiodic induction." Bot. Gaz. 101: 658-687.

Heslop-Harrison, J. 1963. "Sex expression in flowering plants." Brookhaven Symp. Biol. 16: 109-125.

Hicks, G. S., Sussex, I. M. 1971. "Organ regeneration in sterile culture after median bisection of the flower primordia of Nicotiana tabacum." Bot. Gaz. 132: 350-363.

Hillman, W. S. 1975. "Effects of inorganic nitrogen on the response of Lemna carbon dioxide output to light quality and timing." Photochem. Photobiol. 21: 39-47.

Jacobs, W. P. 1972. "Rhythm of leaf development and sensitivity to photoperiodic floral induction." Amer. J. Bot. 59: 437-441.

Jacqmard, A., Miksche, J. P., Bernier, G. 1972. "Quantitative study of nucleic acids and proteins in the shoot apex of Sinapis alba during transition from the vegetative to the reproductive condition." Amer. J. Bot. 59: 714-721.

Jacques, M. 1973. "Transfert par voie de greffage du stimulus photoperiodique." CR Acad. Sci. (Paris) 276: 1705-1708.

Jensen, L. C. W. 1971. "Experimental bisection
    of Aquilegia floral buds cultured in vitro.
    I. The effect on growth, primordia initi-
    ation, and apical regeneration." Canad. J.
    Bot. 49: 487-493.

Katayama, T. C. 1971. "Photoperiodism in the
    genus Oryza." III. Mem. Fac. Agric.
    Kagoshima Univ. 8: 299-320.

Kinet, J. M., Bernier, G., Bodson, M., Jacqmard,
    A. 1973. "Circadian rhythms and the induc-
    tion of flowering in Sinapis alba."
    Pl. Physiol. 51: 598-600.

King, R. W., Cumming, B. G. 1972a. "Rhythms as
    photoperiodic timers in the control of
    flowering in Chenopodium rubrum L." Planta
    103: 281-301.

King, R. W., Cumming, B. G. 1972b. "The role of
    phytochrome in photoperiodic time measure-
    ment and its relation to rhythmic time
    keeping in the control of flowering in
    Chenopodium rubrum." Planta 108: 39-57.

King, R. W., Zeevaart, J. A. D. 1973. "Floral
    stimulus movement in Perilla and flower
    inhibition caused by non-induced leaves."
    Pl. Physiol. 51: 727-738.

Knott, J. E. 1934. "Effect of a localized photo-
    period on spinach." Proc. Amer. Soc. Hort.
    Sci. 31: 152-154.

Kopcewicz, J. 1972a. "Estrogens in the short day
    plants Perilla ocimoides and Chenopodium
    rubrum from under inductive and non-
    inductive light conditions." Z. Pflanzen-
    physiol. 67: 373-376.

Kopcewicz, J. 1972b. "Oestrogens in the long-day
     plants Hyoscyamus niger and Salvia spendens
     grown under inductive and non-inductive
     light conditions." New Phytol. 71: 129-134.

Lang, A., Melchers, G. 1943. "Die photoperiod-
     ische Reaktion von Hyoscyamus niger."
     Planta (Berl) 33: 653-702.

McMillan, C. 1970. "Photoperiod in Xanthium pop-
     ulations from Texas and Mexico." Amer. J.
     Bot. 57: 881-888.

McMillan, C. 1973. "Photoperiod evidence in the
     introduction of Xanthium strumarium to
     Tahiti and Hawaii from Mexico." Amer. J.
     Bot. 60: 277-282.

Moshkov, B. S., Odumanova-Dunayeva, G. A. 1973.
     "The influence of photosynthesis on the
     development of Perilla ocymoides L. and
     Brassica carinata A. Braun under conditions
     of permanent illumination." Botan. Zhur.
     58: 639-645.

Nagl, W. 1973. "Photoperiodic control of activi-
     ty of the suspensor polytene chromosomes in
     Phaseolus vulgaris." Z. Pflanzenphysiol.
     70: 350-357.

Nakayama, S., Borthwick, H. A., Hendricks, S. B.
     1960. "Failure of photoreversible control
     of flowering in Pharbitis nil." Bot. Gaz.
     121: 237-243.

Pol, P. A. van de. 1972. "Floral induction,
     floral hormones and flowering." Meded.
     Landbouwhogesch. Wageningen 72: (9)
     1-89.

Pryce, R. J. 1972. "Gallic acid as a natural
    inhibitor of flowering in Kalanchoë
    blossfeldiana." Phytochem. 11: 1911-1918.

Schwabe, W. 1956. "Evidence for a flowering
    inhibitor produced in long day in Kalanchoë
    blossfeldiana." Ann. Bot. N. S. 20: 1-14.

Schwabe, W. W. 1970. "The control of flowering,
    growth and dormancy in Kleinia articulata
    by photoperiod." Ann. Bot. 34: 29-41.

Schwabe, W. W. 1972. "Flower inhibition in
    Kalanchoë blossfeldiana: bioassay of an
    endogenous long day inhibitor and inhibi-
    tion by (+) abscisic acid and xanthoxin."
    Planta (Berl.) 103: 18-23.

Searle, N. E. 1961. "Persistence and transport
    of flowering stimulus in Xanthium."
    Pl. Physiol. 36: 656-662.

Soetiarto, S. R., Ball, E. 1969. "Ontogenetical
    and experimental studies of the floral apex
    of Portulaca grandiflora. 2. Bisection of
    the meristem in successive stages."
    Can. J. Bot. 47: 1067-1076.

Staden, J. van, Wareing, P. F. 1972. "The effect
    of photoperiod on levels of endogenous
    cytokinins in Xanthium strumarium."
    Physiol. Plantar. 27: 331-337.

Suge, H. 1972. "Inhibition of photoperiodic flo-
    ral induction in Pharbitis nil by ethylene."
    Plant & Cell Physiol. 13: 1031-1038.

Takimoto, A., Hamner, K. C. 1964. "Effect of
    temperature and preconditioning on

photoperiodic response of <u>Pharbitis nil</u>."
<u>Pl. Physiol</u>. 39: 1024-1030.

Vince, D. 1965. "The promoting effect of far red
    light on flowering in the long-day plant
    <u>Lolium temulentum</u>." <u>Physiol. Plantar</u>. 18:
    474-482.

Wagner, E., Cumming, B. G. 1970. "Betacyanin
    accumulation, chlorophyll content, and
    flower initiation in <u>Chenopodium rubrum</u> as
    related to endogenous rhythmicity and
    phytochrome action." <u>Can. J. Bot</u>. 48: 1-18.

Wellensiek, S. J. 1966. "The flower forming
    stimulus in <u>Silene armeria</u> L." <u>Z. Pflanzen-
    physiol</u>. 55: 1-10.

Yoshida, K., Umemura, K., Yoshinaga, K., Oota, Y.
    1967. "Specific RNA from photoperiodically
    induced cotyledons of <u>Pharbitis nil</u>."
    <u>Plant & Cell Physiol</u>. 8: 97-108.

Zeevaart, J. A. D. 1958. "Flower formation as
    studied by grafting." <u>Meded.
    Landbouwhogesch. Wageningen</u>. 58: (3) 1-88.

Zeevaart, J. A. D. 1962. "The juvenile phase in
    <u>Bryophyllum daigremontianum</u>." <u>Planta
    (Berl.)</u> 58: 543-548.

Zeevaart, J. A. D. 1973. "Gibberellin A_{20} con-
    tent of <u>Bryophyllum daigremontianum</u> under
    different photoperiodic conditions as
    determined by gas liquid chromatography."
    <u>Planta (Berl.)</u> 114: 285-288.

Index

119